W9-COJ-072

Sampling Methods for the Auditor: An Advanced Treatment

Sampling Methods for the Auditor

An Advanced Treatment

HERBERT ARKIN

Professor Emeritus
Bernard M. Baruch College
City University of New York

McGRAW-HILL BOOK COMPANY

New York / St. Louis / San Francisco / Auckland
Bogotá / Hamburg / Johannesburg / London / Madrid / Mexico
Montreal / New Delhi / Panama / Paris / São Paulo
Singapore / Sydney / Tokyo / Toronto

Library of Congress Cataloging in Publication Data

Arkin, Herbert, 1906–
Sampling methods for the auditor.

Includes index.
1. Auditing. 2. Sampling (Statistics)
I. Title.
HF5667.A694 657′.45 81-2735
ISBN 0-07-002194-5 AACR2

1234567890 KPKP 8987654321

ISBN 0-07-002194-5

The editors for this book were Ann Gray and Jon Palace,
the designer was Jules Perlmutter,
and the production supervisor was Paul Malchow.
It was set in Melior by University Graphics, Inc.

Printed and bound by The Kingsport Press.

Contents

Preface

The rapidly increasing recognition of the need for the use of statistical (probability) sampling in conjunction with audit examinations has resulted in the education of a large and increasing number of auditors in the application of these techniques.

This educational effort has been given impetus by the recognition of the importance of these techniques by the American Institute of Certified Public Accountants both by the appearance of questions in this area on the CPA examination and the incorporation of material on sampling in auditing in *Statement on Auditing Standards 1.*[1]

Many of the larger firms have included segments on this subject in their orientation courses for "new hires," and some have developed basic and advanced courses in the subject. While there are few courses entirely devoted to probability sampling techniques for auditors at colleges and universities, basic auditing courses now frequently devote a portion of their time to this subject.

Federal, state, and local government agencies have devoted considerable training efforts to this subject. The author has for the past 20 years conducted a basic course in statistical sampling for auditors employed by government agencies, in which thousands of auditors have been enrolled. Other courses in statistical sampling for auditors have been conducted by various accounting societies and by public accounting firms as part of their training efforts.

With the spread of knowledge of the subject has come increased sophistication in its application and a desire for knowledge beyond that normally covered in these elementary courses.

It is the purpose of this book to provide this more advanced knowledge to enable the auditor to deal validly and efficiently with the application of the sampling techniques to audit problems. For this reason, it

[1]*Statement on Auditing Standards 1*, American Institute of Certified Public Accountants, New York, 1973, appendixes A and B.

is assumed that the reader has a knowledge of the basic principles of probability (statistical) sampling.

In addition, coverage is supplied relative to the various practical considerations which are encountered when attempting to apply the statistical sampling methods in the audit situation. The types of difficulties which are frequently encountered are discussed in detail.

Last, but certainly not least, the differences in the basic philosophies as to the application of these methods, including the views expressed by the American Institute of Certified Public Accountants in its publications, the approaches of various public accounting firms, as well as the pronouncements of various authors, are explored in depth. In addition, some possible new approaches are explored.

Sampling Methods for the Auditor: An Advanced Treatment

Chapter 1

Sampling and Audit Objectives

The ultimate objective of the audit examination of financial statements by the independent public accountant is the expression of an opinion on the fairness with which the financial statements present the financial position of the organization audited.

Thus, the short-form report recommended by the American Institute of Certified Public Accountants (AICPA) reads, in part, "in our opinion, the aforementioned financial statements present fairly the financial position of the X Company . . . in conformity with generally accepted accounting principles applied on a basis consistent with that of the prior year."[1]

However, in the Continental Vending case Judge Friendly, citing the opinion of the Circuit Court Judge (*United States v. Simon*, Court of Appeals, Second Circuit 1969), emphasized that "the critical test is not whether the statements were prepared in accordance with generally accepted accounting principles, but whether they fairly present financial information such that they contain no misstatement of fact, or, at least no misstatement of fact known to the auditor."[2] This view places an emphasis on the *validity of the values* composing the financial statements.

A report of a Senate subcommittee (the Metcalf report) states that "the independent auditor is also responsible for certifying the accuracy

[1] *Statement on Auditing Standards 1*, American Institute of Certified Public Accountants, New York, 1973, p. 81.

[2] The opinion of the Court of Appeals as reprinted in the *Journal of Accountancy*, February 1970, p. 65.

of corporate records to the public."[3] This statement further indicates a responsibility for the validity of reported accounting data on the part of the independent public accountant.

The audit examination by an independent public accountant includes a variety of reviews and examinations which are intended to provide the auditor with a sound basis for the expression of an opinion relative to the fairness of the financial statements, as well as a basis for the necessary conclusion that the statements have been prepared in accordance with generally accepted accounting principles.

The AICPA publication *Statement on Auditing Standards 1 (SAS 1)* prescribes that "sufficient evidential matter is to be obtained through inspection, observation, inquiries and confirmations to afford a reasonable basis for an opinion regarding the financial statements under examination."[4]

Confronted with the masses of documentation which develop in modern business, the auditor finds it impractical to examine all of the entries or other records and resorts to the examination of a portion of these documents. This partial examination is called an audit test and, of course, constitutes a sample.

To fully appreciate the attitude of the auditor toward the test (or sample) and its consequent impact on the development of the use of statistical sampling in auditing, it is necessary to consider the history of the audit approach.

THE AUDIT APPROACH AND SAMPLING

In an earlier day, when business organizations were smaller, the auditor could obtain the required assurance as to the accuracy of the records by examining all of the transactions in detail.

Montgomery, writing in 1912, referred to this type of audit as "bookkeeper audits" and described them as consisting of vouching all cash disbursements, checking all footings and postings, checking the ledger to the trial balance and the trial balance to the financial statements.[5]

[3] *Report of Subcommittee on Reports, Accounting and Management of the Committee on Government Operations*, United States Senate, 94th Cong., 2d Sess., December 1976, as reprinted in the *Journal of Accountancy*, March 1977, p. 113. To clarify the reference in this statement, it should be noted that the report then went on to say that "present auditing standards permit an independent auditor to use a great amount of discretion in determining how much testing should be done."

[4] *Statement on Auditing Standards 1*, p. 33. This is the third standard of field work.

[5] R. H. Montgomery, *Auditing Theory and Practice*, 1st ed., The Ronald Press Company, New York, 1912, as cited in C. A. Moyer, "Early Developments in American Auditing," *Accounting Review*, January 1951.

This was in the early British tradition, where the primary audit objective was the detection of fraud.

As business organizations grew, this detailed checking became patently impractical, at least in some situations. In addition, in the early part of this century, the audit objective had changed from detection of fraud to the determination of the fairness of the financial statements.[6] At this point (during the early 1900s) the verification method gradually began to change to an examination of a part of the records or the use of the audit test (or sample). During the 1940s increasing emphasis was placed on the protection provided by the internal control system, and later the evaluation of internal control became a matter of primary emphasis, with decreasing dependence on the audit test.

This change is reflected in the *Statement on Auditing Procedure 1,* which states that "there is to be a proper study and evaluation of the determination of the resultant extent of the tests to which auditing procedures are to be restricted."[7]

This ever-increasing emphasis on the internal control system was accompanied by a decline in emphasis on the test. In part, this may have developed from the realization that the test method, as then applied, based solely on the auditor's judgment and lacking in objectivity, did not really provide the protection desired.

Yet sampling techniques were available, at least during the latter part of the period, which could provide the necessary objectivity and hence the desired protection. These were the techniques of probability sampling which were well developed before the 1940s and available for use in the audit test.

Further, an evaluation of the protective quality of an internal control system is of little value unless accompanied by assurance that there is no important frequency of deviation from the system, deliberate or inadvertent. A system that fails frequently does not provide the necessary protection, even though it may be evaluated as excellent on paper.

In addition, there is always the possibility of material error arising from clerical failures or occurring in areas not adequately protected by the internal control system. It is only by examination of the records that protection against such "leaks" in the internal control system can be provided. This examination need not be a detailed 100% examination but may be based on a sample. These examinations are referred to as *tests of compliance.*[8]

[6] R. G. Brown, "Changing Audit Objectives and Techniques," *Accounting Review,* October 1962. It is to be noted that the historical discussion that follows is based on material in this article.

[7] *Statement on Auditing Standards 1,* p. 81.

[8] *Statement on Auditing Standards 1,* p. 47.

Apart from the question of the possible failure or inadequacy of the internal control system, there are sampling techniques which directly relate to the values that appear in the financial statements. These tests of account balances are sometimes referred to as *substantive tests.*[9]

By the use of variables (dollar value) sampling, it is possible to project an estimate of the value of the account balance which would have been obtained had a 100% audit been accomplished. Thus, for instance, the reasonableness of the stated value obtained from a 100% physical inventory can be established by use of an independent sample accomplished by the auditor, where the auditor counts, prices, and extends the sample items and projects an interval estimate of the audited ("true") value of the inventory. Such an approach has in the past been considered by many auditors as too sophisticated and perhaps as having unfavorable legal implications. Nevertheless, the use of these methods is increasing, and the legal concerns are rapidly disappearing. The legal questions involved in sampling are discussed in Chapter 9.

It is not to be implied, however, that the test (sample), when used in this direct manner, replaces either the attention placed on evaluation of the internal control system or other portions of the audit examination.

As will be seen later, sampling procedures *alone* cannot provide complete protection, although such tests are an essential portion of the audit examination. All portions of the audit examination provide a part of the desired assurance, but there is indeed a question of emphasis.

THE INTERNAL AUDITOR, THE GOVERNMENTAL AUDITOR, AND SAMPLING

The internal auditor and the governmental auditor arrived on the scene later than the independent public auditor. As a result, their techniques generally have followed those of the public accountant, rather than being developed separately.

The development of the idea that the independent public accountant can rely upon the work of the internal auditor to provide additional assurance has helped give rise to a common approach. This attitude has been reinforced by the fact that the independent public accountant may look at internal auditing as an additional portion of the internal control system or through the concept of "cooperative auditing," where the internal auditor performs tests in cooperation with the public

[9]The compliance test is an attributes test since it is devoted to establishing the frequency with which deviations from internal control or other errors or failures occur. The substantive test is a variables (dollar value) test.

accountant. In either event, the tendency will be for the internal auditor to follow approaches similar to those of the public accountant.

Yet the objectives of the internal auditor and the governmental auditor may differ widely from those of the independent public accountant.

The internal auditor usually does not have as a purpose the expression of an opinion as to the fairness of the financial statements. The internal auditor is concerned with protecting the integrity of the records of the company and with the broader function of reviewing the operations of the company, including financial, accounting, and other operations, to assure compliance with established procedures and to provide the basis for improvement in operations.

The governmental auditor often has functions similar to those of both the independent public accountant and the internal auditor, especially when the governmental auditor is established as an independent agency within a department.

The Comptroller General of the United States indicates that the purpose of government audits includes "not only financial and compliance auditing but also auditing the economy, efficiency and the achievement of desired results."[10] The elements of such an audit are defined as follows:

1. *Financial and compliance*—determines (a) whether financial operations are properly conducted, (b) whether the financial reports of the audited entity are presented fairly, and (c) whether the entity has complied with applicable laws and regulations.
2. *Economy and efficiency*—determines whether the entity is managing or utilizing its resources (personnel, property, space and so forth) in an economical and efficient manner and the causes of any inefficiencies or uneconomical features including inadequacies in management information systems, administrative procedures, or organizational structure.
3. *Program results*—determines whether the desired results are being achieved, whether the objectives established by the legislature or other authorizing body are being met, and whether the agency has considered alternatives which might yield desired results at a lower cost.[11]

As a result, the internal auditor and governmental auditor may have functions beyond those of the independent public accountant. However, the increasing use of the "management report" by independent public accountants to highlight failures or deficiencies in the internal

[10] *Standards for Audit of Governmental Organizations, Programs, Activities and Functions*, Comptroller General of the United States, Washington, D.C., 1972, p. 2.

[11] Ibid., p. 2.

control system, as well as the growth of management advisory services within public accounting firms, tends to bring their objectives closer together and will increasingly require them to confront problems similar to those of the internal auditor.

Nevertheless, the objectives of auditing by public accounting firms and by government agencies are different, and their sampling problems are not the same. Some techniques, discussed later in this book, are appropriate for the independent public accountant but unnecessary for all government audits. Similarly, some techniques will be more appropriate for the government audit and will not lead to the fulfillment of the public accountant's objectives.

THE DEVELOPMENT OF THE USE OF STATISTICAL SAMPLING IN AUDITING

The application of statistical (probability) sampling techniques to auditing operations started slowly in the 1950s, but has spread rapidly since then. While by no means universally applied in audit tests, the need for this technique to lend objectivity to the audit test has become widely recognized. Many of the larger public accounting firms have created programs for training their personnel in this area and some have made it a matter of policy to use statistical sampling techniques in lieu of judgment sampling.

The Committee on Statistical Sampling of the AICPA has issued statements concerning the use of sampling in auditing and guidelines for the use of the techniques.[12] Questions on statistical sampling appear routinely on the standard CPA examinations. In addition, the AICPA has published a series of self-study lessons on statistical sampling as a means of disseminating the knowledge necessary to apply statistical sampling procedures.[13]

A recent article describing the results of a survey regarding the impact of the proliferation of lawsuits against CPA firms observes that examples of extensions of audit procedures arising from such suits have included the initial or increased use of statistical sampling.[14]

The application of statistical sampling in government audits has spread rapidly, possibly because of the tremendous and increasing vol-

[12] *Statement on Auditing Standards 1*, Appendixes A and B, pp. 36–54.

[13] *An Auditor's Approach to Statistical Sampling*, American Institute of Certified Public Accountants, New York. This series includes five programmed lesson series on the basic techniques of statistical sampling intended to provide a self-study course.

[14] James Bedingfield, "The Effect of Recent Litigation on Audit Practice," *Journal of Accounting*, May 1974, p. 55.

ume of documentation confronting these auditors, as well as the availability of auditors trained in statistical sampling methods. Several federal agencies have stated that it is a matter of policy that statistical sampling techniques will be used in audit tests where possible.[15]

The Institute of Internal Auditors has shown great interest in statistical sampling methods and has published a manual entitled *Sampling Manual for Auditors.*[16]

Starting with a few scattered articles in the 1930s and 1940s the literature on sampling techniques for auditors has grown rapidly.

Although much has been written about statistical (probability) sampling techniques, very little discussion can be found dealing with the practical problems of applying these techniques in the audit situation. This volume discusses the special considerations involved when applying the generally accepted statistical sampling principles to the unique problems of the auditor.

While a wide variety of sampling techniques are available and are widely used in such fields as survey sampling, market research, statistical quality control, etc., it cannot be assumed that these techniques, so successful in these other areas, may be transferred without modification or without differences in application to the problems of the auditor.

The objectives of the auditor are unique. The difficulties and the problems of applying widely used sampling techniques to auditing are also unique. Those methods and approaches which have been most successful in other fields may not accomplish the auditor's particular objectives.

The area of application of sampling techniques in auditing relates largely, but not exclusively, to the "audit test." This test, at least for the commercial independent auditor and to some degree for other auditors, is part of a fabric of procedures, not all of a test nature, designed to provide the auditor assurance relative to the outcome of the entire audit examination.

It would be rare indeed for the independent public auditor to be willing to base a decision about certification solely on the results of a single sample, no matter how obtained or how large. This indeed separates

[15]For instance the United States Army *Audit Agency Manual,* Sept. 4, 1962, para. 317, amended to June 21, 1968, para. 3, "Policy," states: "It is the established policy of the US Army Audit Agency that statistical sampling techniques will be used whenever our audits encompass tests of selective transactions or items in order to express an opinion regarding the entire field from which the selection was made." The U.S. Army Audit Agency has been a leader in the application of statistical sampling to audits.

[16]*Sampling Manual for Auditors,* The Institute of Internal Auditors Inc., New York, 1970.

the auditor's use of sampling techniques from those employing it in other areas such as public opinion polls, market research, statistical quality control, sampling of scientific data, etc., where there is *total* dependence on the sample result.

Thus, it is *not* proper to assume that the auditor's sampling problems can be solved by merely applying sampling techniques developed for other purposes in the same manner as those who developed these methods for their special purposes.

On the other hand, it must be remembered that the basic principles of probability sampling apply in *all* fields. The techniques are merely an application of these basic principles. Nevertheless, the methods may well be applied in different ways or new techniques based on the same principles may be developed to meet the auditor's objectives.

SAMPLING IN AUDITING VERSUS SAMPLING IN OTHER FIELDS

One outstanding feature of sampling procedures in auditing is the limitation of time, resources, and money, which invariably restricts the sample sizes much more than in other fields.

In many fields, the sample point estimate (the single value projected from the sample) is the value used. This is made possible by using whatever sample size is necessary to produce a sampling error of sufficiently small magnitude as to be considered negligible. In the audit situation, the auditor must make do with an interval estimate rather than a point estimate, since the sampling error may be considerable. Nevertheless, it may be entirely possible to use such interval estimates in lieu of the point estimate with useful results. Or it may be possible to use the confidence limits to achieve the desired results.

For instance, instead of establishing a precise figure for the amount of disallowances, to be used as a basis for a claim, limitations of cost, time, and resources may require the use of the lower confidence limit on the ground that the total disallowance is *at least* that figure. To establish that figure on the basis of a point estimate with an innocuous sampling error might cost more than the potential recovery or might be beyond the capability of the audit organization because of time and resource limitations.

In the compliance test situation, an interval estimate may disclose an unsatisfactory situation without the need for a precise point estimate of the rate of occurrence of failures or errors. An interval estimate of the total audited value of an inventory can be used to establish the reasonableness of the book record value, provided that the total spread of the interval is not more than the amount considered to be material in the accounting sense.

The total amount of certain tax credits taken, based on numerous data, may be challenged if the upper limit of the interval estimate is lower than the reported amount without the need for the sampling error to be very small.

Thus, many of the more traditional approaches are not appropriate in the audit situation and may be replaced by new interpretations. In fact, in certain newly introduced techniques (see Chapter 7), the sample estimate made in the audit situation is not that for the actual amount but rather what it might be, given certain extremely conservative assumptions. Those who have knowledge of the more traditional techniques and interpretations must, while insisting upon the application of valid sampling principles, modify their viewpoints to take into account the new techniques.

In addition, there are profound differences *within* the field of auditing which require varied methods of application and interpretation.

Data from samples taken by public accountants are of concern, almost entirely, only to auditors and no one else. The data provide support and comfort for the auditors' certification, *when considered together* with the results of other parts of the audit examination.

However, the governmental auditor is not ordinarily called upon to provide certifications and in many instances the sample data must stand alone. To a large extent the audit is often an operational or management audit.

Thus, if dissallowances are to be claimed as a result of an audit of a contractor or of a taxpayer by a governmental auditor, only the sample projection can be used, and it must be completely defensible in itself. Again, if an audit of welfare payments is to be made, the number and dollar value of payments deemed improper depend solely on the sample projection.

However, there is an overlap. Governmental auditors may be called upon to establish the reasonableness of a reported book value. Independent public accountants may find it necessary to establish a specific figure, such as the reasonableness of a reserve for bad debts or the amounts to be reported for tax purposes. They may need operational data from samples for a "management letter" or for special projects involving the sampling of accounting data. They may be called upon to audit values on the books of record which have been determined by probability sampling methods, such as those for intercompany settlements based on samples, as described later.

Commercial internal auditors have problems which come close to those of governmental auditors although they may produce data upon which the independent public accountants rely.

SCOPE OF BOOK

The material composing this volume is presented with the assumption that the reader has a background in the principles of sampling and preferably has read at least one of the books describing the statistical sampling principles and techniques appropriate to auditing.[17]

Prior study is essential to understanding the material contained in this volume. Readers are cautioned that without such prior knowledge, they will find the material discussed either difficult or obscure and may misconstrue the points made.

It is the objective of this volume to explore the problems of applying statistical sampling in auditing in greater depth than previously attempted. The exploration will be from the viewpoint of both the independent public and the governmental auditor with their separate objectives dealt with individually.

Based on the assumption of a basic knowledge of the statistical sampling technique, the methods are reexamined with respect to the unique problem of using the techniques in the audit situation and the implications arising from these applications.

[17]Those readers who require further background knowledge relative to statistical sampling techniques or find it necessary to refresh their memories in this area are referred to H. Arkin, *Handbook of Sampling for Auditing and Accounting*, 2d ed., McGraw-Hill Book Company, New York, 1974.

Chapter 2

Compliance Tests

As defined by the AICPA, the purpose of tests of compliance is to provide reasonable assurance that the internal control procedures are being applied as prescribed.[1]

More broadly, the objective of a compliance test is to establish the *frequency* of failure to comply with a system, with regulations, or with the law, or the frequency of mistakes of any kind. Obviously, this is a form of attributes sampling and does not relate to dollar values.

The ultimate purpose of such a test is to establish the extent of procedural errors, which may or may not have a dollar value impact. However, the existence of such errors, indicating a procedural breakdown, may, if it has not done so already, result in a future failure of the system with resulting possible material errors.

The function of compliance tests as used by the independent public accountant is to establish whether the internal control system being evaluated is indeed operative and, thus, to provide assurance against the possibility of a material error in the financial statements. It is part of a complex of analyses and tests composing the audit examination.

The objective of compliance tests for governmental auditors (or other internal auditors) is likely to be broader, since it usually includes determining whether the audited entity complied with applicable laws and regulations and whether the financial operations were properly conducted.

The different approaches available for applying statistical sampling in a compliance test will not necessarily be equally appropriate for both groups of auditors. For this reason, the subject of the compliance test is dealt with separately for the two classes of auditors.

[1] *Statement on Auditing Standards 1*, p. 28.

COMPLIANCE TESTS AND THE INDEPENDENT PUBLIC ACCOUNTANT

The publications of the AICPA indicate that the second standard of field work for audits by independent public accountants is that "there is to be a proper study and evaluation of the existing internal control as a basis for reliance thereon and for the determination of the resultant extent of tests to which auditing procedures are to be restricted."[2]

The primary purpose of the prescribed study and evaluation of internal control is to provide the public accountant further assurance to support an eventual statement of opinion relative to the financial statements. However, the AICPA has now added an additional requirement *Statement on Auditing Standards 20,* which provides that "the auditor communicate to senior management and the board of directors or its audit committee material weaknesses in internal accounting control that have come to [the auditor's] attention during an examination of financial statements made in accordance with generally accepted auditing standards."[3]

That publication provides that such weaknesses determined by the review of the system *and* disclosed by tests of compliance will be so reported.

It is emphasized that the above study of internal control includes two phases: "(a) knowledge and understanding of the procedures and methods prescribed and (b) a reasonable degree of assurance that they are in use and are operating as planned."[4]

The second phase indicated above is effectuated by use of "compliance" tests. These tests are ordinarily based on the examination of documentary evidence of compliance on a partial or sample basis. This is obviously a fruitful area for the application of statistical sampling.

It is important to emphasize that the AICPA defines the internal control system broadly as comprising "the plan of organization and all of the coordinate methods and measures adopted within a business to safeguard its assets, check the accuracy and reliability of its accounting data, promote operational efficiency, and encourage adherence to prescribed managerial policies."[5]

The necessary assurance that the internal control system is actually

[2]Ibid., p. 13.

[3]*Statement on Auditing Standards 20, Required Communication of Material Weaknesses in Internal Accounting Control,* American Institute of Certified Public Accountants, New York, August 1977 (effective Dec. 24, 1977).

[4]*Statement on Auditing Standards 1,* p. 27.

[5]Ibid., p. 15.

in satisfactory operation is achieved through tests of compliance (sometimes called tests of transactions). A series of such tests may be required to establish that the internal control system, *in operation*, actually does provide protection.

The objectives of such tests would be to establish whether there are a sufficient number of instances where the internal control system was evaded (deliberately or accidentally), or where other deviations from the system occurred, to negate the effectiveness of the internal controls. In addition, information may be sought on the frequency of clerical errors or mechanical failures in the company operations or records, which may have escaped the internal control system.

Such a test is not to be confused with the traditional "walk-through" where the auditor follows through one or two transactions to establish that the system is as described to him. A walk-through is not a substitute for a test of compliance. A walk-through does *not* establish the frequency of failures to comply with the system. It is used to establish that the system and accompanying procedures are as described.[6]

In any large system, no matter how sound the internal controls may be, there are likely to be at least a few failures of or deviations from the system, if from no other cause than clerical error. Other inadvertent deviations may arise from carelessness or lack of knowledge.

Of course, if the deviations found in the sample are such that none can be tolerated, such as the discovery of an instance of fraud, the mere detection of one instance will bring about appropriate action as soon as it is located. However, in other than such situations, a single error due to human failure may not be considered sufficient to negate the protective features of the internal control system and permit material error.

If these failures are negligible in number and indicate no deliberate avoidance of or basic defect in the internal control system, they will probably be tolerated as arising from human fallibility, especially where the cost of eliminating these infrequent errors would be appreciable.

However, a conclusion must be reached as to the significance of the actual frequency of such occurrences. While it may be desirable to establish the dollar value of such errors, if they are purely procedural in nature this may not be possible. While the lack of proper authorizing signatures or the lack of support documents *in the instances in the test sample* may not have any dollar value effect, it is to be recognized that

[6] *Statement on Auditing Procedure 54*, American Institute of Certified Public Accountants, New York, November 1972, p. 248, states that with respect to the "practice of tracing one or a few of the different types of transactions involved . . . this practice may be useful for the purpose indicated and may be considered as *part* of the tests of compliance . . ." (emphasis added).

there may be other such instances in the population which do have a dollar value impact.

SAS 1 states that "evaluation of precision in terms of frequency of deviation from internal control procedures or other errors not evaluated in monetary terms contributes to the auditor's ultimate purpose by influencing [a] judgment as to the reliability of the records and the likelihood of errors having a monetary effect."[7]

Even for errors which do have a dollar value impact, and for which a projection can be made relative to the total dollar value of such errors in the population, the mere existence of such failures relates to the state of compliance with internal controls. Frequent but small dollar value evasions of the internal control system, which project to an amount less than a material error, may indicate an extensive failure of internal control with the possibility of a rare single deviation of material amount not included in the sample. In any case, the frequent occurrence of such events with large, small, or no dollar values indicates a leak in the internal control system with which the auditor must be concerned.

If it were possible to audit every transaction and every record, the rate of occurrence of the events could be established exactly. Thus, the auditor could conclude upon completion of such an examination that, say, 2.766% of the documents requiring a certain authorizing signature did not have such a signature. Or in another instance, 3.157% of the invoices were incorrectly priced.

THE MAXIMUM TOLERABLE ERROR RATE

With these precise figures in hand, the auditor would then be required to come to a decision. Is the error rate, as indicated by those percentages, sufficiently high to require some further action on the auditor's part? Does it indicate a breakdown in the internal control system? Should tests and other audit procedures be extended? Should management be notified of leaks in the internal control system with a need for correction? Or possibly, if a situation which is very dubious is indicated, should the auditor even consider qualifying the certification?

The auditor's decision must take into consideration the *number* of errors in the population as well as the *rate* of occurrence. If errors are numerous, they are more dangerous than if they are few.

Thus, the purpose of such examinations, performed on a 100% basis or on a sample basis, is to establish whether or not the rate of error or number of errors in the population is sufficiently high to require further action. It is the auditor's function to exercise judgment at this point as to whether some action is required. The auditor may decide to extend

[7] *Statement on Auditing Standards 1*, p. 38.

the audit procedures, place greater emphasis on tests of account balances, or take whatever other steps the situation might dictate.

When judgmental sampling is used the auditor does not compare the rate of deviations found in the sample to some level of errors considered to be critical, determined either before or after the examination, but merely considers whether or not the errors in the sample indicate that the operation of the internal control system has been seriously impaired.[8] If so, the auditor will take appropriate action.

The approach based on specifying a level of errors which would precipitate action has come into wider use with the more quantitative approach arising from the application of statistical sampling to audit tests. The level of failures, which, if it exists, requires further action, will be referred to as the *maximum tolerable error rate* (MTER).

It is to be noted that the MTER does *not* imply a precise figure, so that an auditor confronted with 3.67% of error may conclude that no further action is required, but if the error rate is 3.68%, must decide that something has to be done. In other words, the MTER must be looked upon as an area or zone, and will indicate that a particular rate, say, one *in the vicinity of* 5% or higher, means that further action must be taken.

The MTER will vary with each kind of deviation or error, so that within an audit of a given internal control system for different kinds of failures, several MTERs may be used.

Burns and Loebbeke list four factors which they consider to be especially significant in establishing the MTER. These factors are:

1. The relative and individual importance of pertinent procedures
2. The quantitative aspects of potential procedural deviations
3. The potential natures and amounts of any monetary errors or irregularities believed possible
4. The degree of reliance planned for the controls in question[9]

[8]It is assumed that whether the sample is a statistical or judgment sample, each deviation found will be examined in relation to its import regardless of its frequency or dollar value.

Statement on Auditing Standards 1 states that "in addition to the statistical evaluation, the auditor should also consider the nature and cause of errors revealed by the sample and their possible relation to other phases of [the] examination" (p. 39).

Further, it is noted that "in addition to statistical evaluation of the quantitative significance of deviations from the pertinent procedures, consideration should be given to the qualitative aspects of these deviations. These include (a) the nature and cause of errors, such as whether they are errors in principle or in application, are deliberate or unintentional, are due to misunderstanding of instructions or careless compliance, and the like and (b) the possible relationship of errors to other phases of the audit" (p. 47).

[9]D. C. Burns, and J. K. Loebbecke, "Internal Control Evaluation: How the Computer Can Help," *Journal of Accountancy*, August 1975.

Some types of errors, such as fraud, defalcation, deliberate evasions of internal control, etc., are intolerable and hence result in an MTER of zero percent. In that case discovery sampling is the proper approach rather than the attributes estimation sampling techniques described in this chapter.[10]

THE TWO-SIDED APPROACH

When a sample is used in lieu of a 100% test, the sample estimate replaces the exact figure for the population. Since there is a sampling error which must be recognized, an interval estimate replaces the exact figure obtainable only from a 100% audit.[11] Thus, instead of being supplied with the information that the rate of occurrence in the population is, say, 2.766%, the auditor now knows only that it is between, say, 1.2 and 3.6%. If the interval estimate is sufficiently narrow to provide a conclusion that the population rate of occurrence is above (or below) the MTER, the objective is achieved. This method of evaluating the results of a compliance test sample result (attributes estimation sample) is called the two-sided confidence interval approach. The one-sided approach will be discussed in detail later.

If the confidence interval projection resulting from the sample clearly fixes the population error rate (both upper and lower limits) to be below the MTER, the auditor may decide that no further action is required. If the confidence interval (both upper and lower limits) is above the MTER, some further action on the auditor's part is necessary. If the confidence interval straddles the MTER, with the upper limit above that rate and the lower limit below, the results of the test are inconclusive.

For instance, assume that an auditor decides that if a certain error rate is *in the vicinity* of 5% or is higher, further action is necessary, since this indicates a serious failure of the internal control system. Assume that a sample is obtained which after evaluation produces an interval estimate with a lower limit of 0.4% and an upper limit of 2.8%. The auditor might under these circumstances consider that no further

[10]See H. Arkin, *Handbook of Sampling for Auditing and Accounting*, 2d. ed., McGraw-Hill Book Company, New York, 1974, chap. 28, for a description of this method. Discovery sampling provides the sample size necessary to provide a stated probability of finding at least one example of the specified occurrence in the sample if it occurs at some stated minimum rate in the population.

[11]It is to be emphasized that the point estimate (rate of occurrence in the sample used as an estimate of the population rate of occurrence) may be used *only* when the sampling error is negligible. With the sample sizes commonly possible in an audit situation this circumstance is very unlikely.

action is required since the extent of failure in the internal controls with respect to this kind of error is not sufficient to cause concern. Both upper and lower limits are less than the MTER, which is about 5%. This will be referred to as condition type A.

On the other hand, if the confidence interval had been from 5.2 to 8.3%, the auditor must conclude that there definitely is an excessive rate of error and further action is required. This will be referred to as condition type B.

Finally, assume that the confidence interval is from 2.1 to 6.9%. Since the lower limit is less than the MTER and the upper limit is higher than the MTER, no conclusion can be reached as a result of the test. The test did *not* accomplish its objective. This result will be referred to as condition type C.

Thus, to be able to draw a conclusion one of two conditions (type A or B) must exist. The first is that the upper limit (actually both upper and lower) is less than the MTER, indicating a tolerable condition (type A). In the second, the lower limit of the confidence interval estimate (actually both upper and lower limits) is above the MTER, indicating an unsatisfactory condition (type B).

The reason for the failure of the test (condition type C), and an inability to come to a conclusion, may be either the use of a sample which is too small *or* the occurrence of a marginal rate of error in the field.

Of course if the condition is marginal, say a 4.6% error rate in the field, where an MTER of 5% is specified, no reasonable sample size will produce useful results, but then neither would a 100% sample, which would produce only the knowledge that the condition is marginal. No formal conclusion can be obtained. More frequently this inconclusive result arises from the use of too small a sample.

There are considerable economic and time pressures on the auditor. These pressures have resulted, where statistical sampling has not been used, in inordinately small samples in many cases. Because these samples were judgment samples and could not be evaluated in terms of a valid projection, the uselessness of such samples was not apparent. Many samples of 30 units or fewer have been used in audit tests, where in fact no matter how taken, as a judgment or statistical sample, such sample sizes cannot be justified. They will not produce results which would meet the objectives of the compliance test unless the error rate is catastrophically bad.

The sample size used in a test must be large enough to accomplish the test purpose or the test is completely wasted. There is no protection provided by an inconclusive test!

The relationship of the size of the test to its effectiveness may be determined as illustrated below for a statistical sample. Assume that an auditor wishes to establish the frequency of failure to conform to system requirements, with respect to some specific form of deviation. A maximum tolerable error rate (MTER) of, say, 7% is established for purposes of this test.[12] Further, assume that a sample of 100 units is used, to be evaluated at the 95% confidence level (two-sided).

It is to be noted that in order to conclude that the rate of occurrence in the population is *below* the MTER, in this case, both the upper and lower limits of the confidence interval established on completion of the test must be lower than 7%.

Reference to tables for appraisal of attributes estimation samples[13] using the 95% confidence level for a very large population indicates that a 2% rate of occurrence *in the sample* of 100 will result in a confidence interval of 0.2 to 7.0%. Thus, it will be necessary for the sample to contain 2% or less, or for this sample of 100, two or fewer errors, in order to conclude that the rate of occurrence in the population is lower than 7%.

It is now possible to calculate the probability that a sample of 100 containing two or fewer errors would be produced if the population

[12]The selection of a specific value of 7% as the MTER does not imply that this is a *precise* figure. It indicates merely a general level of error. However, for purposes of the following calculations, a specific figure was required.

[13]H. Arkin, op. cit., p. 377.

Table 2-1 **Probability of Two or Less in a Sample of 100 from a Population with Various Rates of Occurrence**

Actual Population Rate of Occurrence, %	Probability of Two or Less in Sample,* %	Probability of Correct Decision,† %
0.5	98.6	98.6
1.0	92.1	92.1
2.0	67.7	67.7
3.0	42.0	42.0
4.0	23.2	23.2
5.0	11.8	11.8
6.0	5.7	5.7
7.0	2.6‡	
8.0	1.1‡	
9.0	0.5‡	
10.0	0.2‡	

*These binomial probabilities can be found by reference to such tables as *Tables of the Cumulative Binomial Probability Distribution*, Harvard University Press, Cambridge, 1955.
†Correct conclusion is that the population contains less than 7% of such occurrences.
‡Incorrect decision.

Table 2-2 ***Probability of 13 or More in a Sample of 100 from an Infinite Population with Various Rates of Occurrence***

Actual Population Rate of Occurrence, %	Probability of 13 or More in Sample, %	Probability of Correct Decision,* %
5.0	0.2†	
6.0	0.7†	
7.0	2.2	2.2
8.0	5.6	5.6
9.0	11.4	11.4
10.0	19.8	19.8
11.0	30.5	30.5
12.0	42.4	42.4
13.0	54.5	54.5
14.0	65.7	65.7
15.0	75.3	75.3
16.0	83.0	83.0
17.0	88.8	88.8
18.0	92.9	92.9
19.0	95.7	95.7
20.0	97.5	97.5
21.0	98.6	98.6

*Correct conclusion is that population contains 7% or more of such occurrences.
†Incorrect decision.

actually contained various rates of occurrence. If the population contains not more than 7% of errors, the correct decision will be reached if two or fewer are contained in the sample since the limits produced are 0.2 to 7.0% or lower. These probabilities are shown in Table 2-1 for various rates of occurrence in a population. On the other hand, *both* the lower and upper limits must be *above* the MTER to conclude that the rate of occurrence in the population is in excess of 7%.

Reference to a table for the evaluation of attributes estimation sampling[14] discloses that a sample of 100 units containing a 13% rate of occurrence of errors will produce a confidence interval of 7.2 to 21.2% with a 95% confidence level. A sample of 100 with 13 or more occurrences would demonstrate the rate of occurrence in the population to be in excess of 7%. The probability that such a sample containing 13 or more instances will be produced when the population contains various rates of occurrence is shown in Table 2-2.

A sample containing two or fewer will indicate a rate of occurrence in the population of less than 7%. A sample containing 13 or more will

[14]H. Arkin, op. cit., p. 410.

indicate a rate of occurrence in the population of 7% or more. However, a sample containing 3 to 12 instances will be inconclusive since the lower limit will be below the MTER (7%) and the upper limit will be above 7%.

The probability of an inconclusive and therefore useless result for the above test for various rates of occurrence in the population is given in Table 2-3.

An audit test serves no purpose unless a useful conclusion can be reached. The probability in the test situation described above that a test sample of various sizes will not produce useful results has been calculated and is shown in Table 2-4. It is to be recalled that the MTER in this example was 7%, and a 95% confidence level was chosen. It is clear from this analysis that a test size of 50 units is of little or no value for this purpose. In order to have even a 50% probability of a *useful* test result, the actual rate of occurrence in the population would have to be less than 2% or in excess of 15%. In other words, when the actual error

Table 2-3 ***Probability of 3 to 12 Occurrences in a Sample of 100 from an Infinite Population Containing Various Rates of Occurrence***

Actual Population Rate of Occurrence, %	Probability of Inconclusive Test Result, %
0.5	1.4
1.0	7.9
2.0	32.3
3.0	58.0
4.0	76.6
5.0	88.0
6.0	93.6
7.0	95.2
8.0	93.3
9.0	88.1
10.0	80.0
11.0	69.5
12.0	57.6
13.0	45.5
14.0	34.3
15.0	24.7
16.0	17.0
17.0	11.2
18.0	7.1
19.0	4.3
20.0	2.5
21.0	1.4

Table 2-4 ***Probability of Inconclusive Test Results for Samples of Various Sizes for Various Actual Error Rates when MTER Is 7% (Infinite Population)***
(95% Two-Sided Confidence Level)

Actual Population Rate of Occurrence, %	Probability in percent of Inconclusive Test Results with Sample Sizes of		
	50	100	200
0.5	22.2%	1.4%	*
1.0	39.5	7.9	0.1%
2.0	63.6	32.3	4.9
3.0	78.2	38.0	25.4
4.0	86.9	76.6	55.0
5.0	92.0	88.0	78.7
6.0	94.5	93.6	91.2
7.0	95.1	95.2	96.5
8.0	94.1	93.3	96.8
9.0	91.4	88.1	80.7
10.0	87.3	80.0	64.8
11.0	81.8	69.5	46.7
12.0	75.2	57.6	30.0
13.0	67.7	45.5	17.3
14.0	59.9	34.3	8.9
15.0	51.8	24.7	4.2
16.0	44.0	17.0	1.7
17.0	36.7	11.2	0.7
18.0	30.0	7.1	0.2
19.0	24.2	4.3	0.1
20.0	19.0	2.5	0.1
21.0	14.8	1.4	*

*Less than 0.1%.

rate in the population is between 2 and 15%, from 50 to 95% of the tests performed with this sample size would be inconclusive and therefore useless, while for often used smaller sample sizes the situation would be much worse!

Yet this is the range within which errors of the type which might have an MTER of 7% are likely to occur. On the other hand, with a sample size of 200, the range of actual rates of occurrence for which there is at least a 50% probability that the sample would not be useful would be from about 4 to 10%, as compared with a range of 2 to 15% for a sample of 50. Even larger sample sizes would be required to narrow the range of inconclusive tests to a more reasonable level.

It is to be noted that this two-sided approach, unlike the one-sided approach discussed below, provides an ability to determine *both* whether the situation is acceptable (condition type A) or unacceptable (condition type B).

THE ONE-SIDED APPROACH—THE AICPA METHOD

Another method for the evaluation of the sample projection is to use only one confidence limit of the interval estimate. The statement may be made that the error rate does not exceed some specified level of errors or alternatively (but not at the same time) that it is lower than some indicated percentage. This is called the one-sided approach.

Statement on Auditing Procedures 54 states that "the auditor's evaluation of compliance would include a statistical conclusion that the procedural deviations in the population did not exceed the upper precision limit obtained or, alternatively, were within the precision range obtained."[15]

The AICPA then uses examples of audit tests which deal with upper limits only and are, therefore, one-sided tests.[16] While, unquestionably, one-sided tests constitute a valid statistical approach, some question may be raised as to their usefulness and the protection they provide in the audit test situation. Much depends on the manner in which the one-sided limits are used.

One method is to use one-sided confidence limits by equating the one-sided upper limit to the MTER and using a sample of a given size to establish the maximum percent (or number) of errors that could

[15] *Statement on Auditing Procedures 54*, American Institute of Certified Public Accountants, New York, November 1972, p. 270.

[16] It is to be noted that this is the approach suggested in the AICPA's programmed lesson series. See "Sampling for Attributes," *An Auditor's Approach to Statistical Sampling*, vol. 2, American Institute of Certified Public Accountants, New York, 1974, p. 26, frame 1-25, and answer on p. 27. However, it is to be noted in *Field Manual for Statistical Sampling*, vol. 6, American Institute of Certified Public Accountants, p. 11, of the same series the observation is made "When the auditor's objective is to estimate the population occurrence rate, [the auditor will] calculate a precision interval and utilize both precision limits. When [the] wish is to decide whether the population occurrence rate exceeds some tolerable limit, an estimate [will be made] and concern [taken] only with the upper precision limit."

It must be emphasized that the use of only the upper precision limit makes it impossible to prove that "The population occurrence rate exceeds some tolerable limit," only that it is *lower* than that limit. Yet failure to demonstrate that the upper limit is lower than the MTER does not prove that the actual population rate is above the MTER.

occur in the sample and still allow the auditor to consider the error rate in the population satisfactory.

The sample size to be used in this approach is established by specifying an expected rate of occurrence, an MTER, and a one-sided confidence level. A frequently selected expected rate of occurrence is zero, that is, that no errors will be found in the sample. By reference to appropriate tables[17] a sample size which will yield an upper limit equal to the MTER, if the expected number of occurrences are found in the sample, may be determined. Then if the number of errors found in the sample is the number expected or less, the population occurrence rate most probably (with probability equal to the confidence level) is less than the MTER.

For instance, a popular approach, when the MTER is 5%, is to use a sample size of 60, which would result in a one-sided upper limit of 4.9% at the 95% confidence level, if no errors are found in the sample for a large population. This is a sampling plan often selected for the one-sided method.

If no errors are found in the sample, the error rate in the population is considered to be acceptable, since there is a 95% probability that the population error rate is lower than 5%. Such a conclusion is valid and will complete the test of compliance.

However, it is to be emphasized that when a one-sided upper confidence limit is used in this manner, there is no way by which the population can be declared unsatisfactory because of the occurrence of more errors than that specified by the MTER. To be able to state that the population is unsatisfactory, the *lower* confidence limit must be established to be *above* the MTER. Since the lower limit is not considered in this method, such a determination is not possible. In other words, the method seems to be devised to prove the population to be satisfactory and provides no valid alternative.

The mere fact that some errors are found in the sample is not conclusive proof that the population is unacceptable. It may be acceptable *or* unacceptable when some errors are found in the sample.

It is to be noted that there is no requirement that the plan used be based on the finding of no errors in the sample, as in the above example. For instance, a plan with a sample size of 100 units with a requirement of a 1% error rate (or one error) in the sample provides a one-sided upper confidence limit of 4.7% and could be used for the same purpose.

[17]Tables for this purpose include the appendix F tables in H. Arkin, op. cit., with one-sided confidence levels, or the tables in "Supplementary Section, Sampling for Attributes," *An Auditor's Approach to Statistical Sampling*, pp. 5-15.

However, a question can be raised as to the likelihood of a correct conclusion and the accomplishment of the test's purpose when the error rate is at various levels below the MTER for a good (low-error-rate) population. Further, a problem arises when more errors are found than the number allowed for a conclusion that the population is satisfactory. In this case one error would prevent such a conclusion.

To be useful such a plan must have a high probability of leading to a correct conclusion that the error rate is satisfactory when the rate of occurrence in the population is below 5% for the above example. If a sample of 60 units is taken, the probability that there will be zero errors in the sample when the population contains various levels of error below 5% then becomes the probability that the plan will lead to the correct conclusion. (See Table 2-5.)

It will be seen from Table 2-5 that the probability that a correct conclusion can be reached that the error condition of the population is satisfactory (below the MTER of 5%), and therefore that the test can be terminated without further action, is very small unless the population contains no errors or virtually none.

It will be recalled that the MTER in this example is 5%. Any population with a lower error rate should be accepted as satisfactory. Yet if the error rate in the population is as low as 1%, there is only a 54.7%

Table 2-5 ***Probability of Correct Conclusion that Population Contains Less Than the MTER of 5%, with a Sample of 60 Containing No Errors for a Large Population (One-Sided Approach)***

Actual Population Error Rate, %	Probability of No Errors in Sample of 60,* %
0.0	100.0
0.5	74.0
1.0	54.7
1.5	40.4
2.0	29.8
2.5	21.9
3.0	16.1
3.5	11.8
4.0	8.6
4.5	6.3

*Leading to correct conclusion that population has a rate of occurrence less than the MTER (in this case 5%).

probability that the population will be declared satisfactory and no further action taken. A 1% rate of occurrence is far from the MTER of 5% but has only about a 50-50 probability of being declared satisfactory. With a 2% rate of occurrence of error, the probability drops to 29.8%, or less than 1 chance in 3.

Many choose a sampling plan of this type because of the small sample sizes associated with the plan. Yet unless the plan is able to declare populations which are satisfactory to be acceptable, the test is a waste. Further action would then be required unnecessarily. The savings anticipated by the use of a small sample size are likely to be an illusion, unless the population is nearly perfect. Even rates of error well below the MTER are likely to cause the test to fail.

When the one-sided approach is used and the number of errors found in the sample is too large to assure that the upper confidence limit is below the MTER, some further action is taken.

Alternative actions may be considered in this situation. One possibility is to assume that the population is unsatisfactory and to take the steps that are necessary when it is not possible to declare the error rate satisfactory. The most probable actions include widening the audit scope, and vigorously testing the account balances.

It is apparent that if this approach is used, unless the population is perfect or very close to perfect, the probability of false rejection and of performing unnecessary additional audit work is very high. As noted, this arises out of the probability that a perfectly satisfactory population often will be rejected.

Thus, the apparent initial saving resulting from a small sample size will, on the overall, result in a much larger audit expense over a series of audits. In many of the compliance tests, errors will be found beyond those permitted by the sampling plan with additional audit effort resulting because of the false alarm, even though the error rate is actually much lower than the MTER.

In fact, when the sample contains some errors, since only an upper limit has been established, the only valid conclusion that can be drawn is that the rate of occurrence is lower than some figure which in turn is higher than the MTER. Of course, reappraisal of the sample will produce a new, higher, *upper* limit but again the only possible conclusion is that the rate is probably lower than this new higher figure which in turn is higher than the MTER. A not very useful projection!

Because of the smaller sample sizes associated with this technique, it is quite likely even if this evaluation were done on a two-sided basis that the interval estimate would be so wide as to be of questionable value. For instance, where a sample of 60 was used in this approach, and where an error was found in the sample (1.7%), the two-sided confidence interval at the 95% confidence level is from 0.1 to 9.0%. Recall that the MTER is 5% in this case.

However, the second alternative is the one most commonly used. When the auditor finds *too many errors* in the sample to conclude that the upper limit is not higher than the MTER, the temptation is to add to his initial sample.[18] After this additional sample has been taken, he then attempts to draw a conclusion based on the results of both samples combined. The sampler then resorts to the usual tables for evaluation of attributes samples to establish a new upper limit.[19]

Unfortunately, these tables (or values obtained from the formulas on which these tables are based) will no longer produce correct confidence limits and may result in considerable underestimates of the actual confidence limits.

EXPANDING THE SAMPLE—TWO-STEP SAMPLING

When an attributes estimation sample is taken in two steps, with an evaluation of the initial sample intervening, the confidence intervals or limits obtained from the traditional tables may not apply, depending on the basis for the sample extension.

The most usual reason for expanding the sample is an inability to declare the population satisfactory on the basis of the initial sample. (If the population can be declared to be satisfactory, no further action is required by the auditor.) If this inability is the reason for the enlargement of the sample after evaluation of the smaller initial sample, the confidence limits from the traditional tables no longer apply. This condition is true whether the test is performed on a one-sided *or* two-sided basis.

The failure to provide accurate confidence limits by the usual method under such circumstances for a two-step sample may be established

[18]This is the method suggested in the AICPA publication, "Sampling for Attributes," *An Auditor's Approach to Statistical Sampling*, vol. 2, p. 20, frame 3-33, and vol. 6, *Field Manual for Statistical Sampling*, p. 20. This suggested approach, however, is contradicted in vol. 6, *Field Manual for Statistical Sampling*, of the same series, where it is stated that "*A possible reaction to this kind of result is to consider expanding the sample as it relates to attributes, in hopes that the upper precision limit will be reduced to an acceptable level. . . . This approach is not advisable. . . . Chances of obtaining no additional errors are extremely rare*" (p. 20).

[19]Tables may be obtained from H. Arkin, op. cit., appendix F, or "Supplementary Section, Sampling for Attributes," *An Auditor's Approach to Statistical Sampling*, table 2, exhibit II.

both on the basis of general logic and more specifically on the basis of statistical theory.

Examination of the logic of such a procedure will quickly demonstrate the defect in the reasoning. The auditor, in approaching the test in this fashion, in effect has decided that if the population cannot be declared to be satisfactory on the basis of the initial sample then *another try will be made to prove it satisfactory.* Since there is always a risk in any form of sampling, theoretically the auditor could continue this process of incrementing the sample size indefinitely, until due to this risk the conclusion could be reached that the population was satisfactory. This repeated effort toward proving a satisfactory condition obviously is not an objective approach.

Further, an examination of the statistical theory involved indicates that the assumption that a two-step sample is equivalent to a single sample of the size obtained by adding both sample sizes is *not* correct.[20]

To illustrate this situation, the example used above for a one-sided test may be examined. In the example cited, a sample size of 60 units is to be used on a one-sided confidence limit basis at the 95% confidence level for a large population and with an MTER of 5%. If no errors are found, the population will be considered satisfactory and the test terminated. If any errors are found the sample will be enlarged by taking an additional 140 sampling units which will be included in the sample.

The basis for establishing the confidence interval estimate involves the concept of the sampling distribution. This is the tabulation of the results (in this case the percentage of errors in each sample) that would be obtained if *all possible* samples of a given size were drawn from a given population containing errors at some specified rate.

The resulting sampling distribution for a single sample of 200 units taken from a large population containing 5% of some type of error is given in Table 2-6. Assume that a two-step compliance test is to be performed on such a population. The auditor decides to draw a preliminary sample of 60 units. If no errors are found in the sample, the test

[20]This type of sampling is similar to double sampling or, when several sample increments are involved, multiple sampling, as applied in statistical quality control. The plan described here, however, differs from the traditional double sampling plan as used in statistical quality control in that traditional plans in this area prescribe both an accept and reject number with the sample increment taken only if the number in the sample falls between the two values. In the plan usually used by auditors, the sampler can only accept on the basis of the first sample or extend the sample. There is no basis for rejection in the first sample. The audit philosophy is that if the population cannot be established as acceptable, the test should be expanded. A further difference is that the acceptance sampling approach does not provide a confidence interval estimate but merely an accept-reject decision.

Table 2-6 ***Sampling Distribution of a Percentage for Sample of 200 from a Large Population with a 5% Rate of Error***

Number of Errors in Sample	Percent of Samples
0	*
1	*
2	0.2
3	0.7
4	1.7
5	3.6
6	6.1
7	9.0
8	11.4
9	12.8
10	12.8
11	11.7
12	9.7
13	7.4
14	5.2
15	3.4
16	2.1
17	1.2
18	0.6
19	0.3
20	0.1

*Less than 0.1%.

will be terminated and compliance with internal control will be considered satisfactory.[21] If, and only if, an error is found, an additional sample of 140 units will be taken. The auditor will then establish a confidence interval estimate for the entire 200 sampling units included in the test by reference to an appropriate table. Let

$$n_1 = \text{initial sample size}$$
$$n_2 = \text{additional sample size}$$
$$n = \text{combined sample size}$$

[21]This is equivalent to considering the MTER equal to 5%, if the 95% confidence level is used on a one-sided basis. See H. Arkin, op cit., table F27, where an upper limit is given as 4.9% for a population of 100,000 and over when the sample size is 60 units and no errors are found in the sample.

It is evident that the incremental sample will be taken *only* if some errors are found in the *first* (initial) sample. With a 5% rate of occurrence of error in the population and a sample of 60 units, it can be calculated that 4.6% of all possible samples will contain no errors and thus will *not* require the additional 140 sampling units. For these 4.6% of all samples there will be no second (additional) sample of 140 units. Hence the sampling distribution will be different. The sampling distribution for such a situation for this two-step sampling plan as contrasted with a single sample of 200 units taken all at once is shown in Table 2-7.

Since the sampling distribution used to establish the confidence interval estimate for a two-step sample (in this case of 60 units and an

Table 2-7 **Sampling Distribution of Percentage for Single- and Two-Step Sample of 200, When p = 5%, for a Large Population**

	Percent of Samples	
Number of Errors in Sample	Single Sample of 200	Two-Step Sample 60 and 140*
0	†	4.6
1	†	†
2	0.2	0.1
3	0.7	0.4
4	1.7	1.3
5	3.6	3.0
6	6.1	5.4
7	9.0	8.2
8	11.3	10.8
9	12.8	12.3
10	12.8	12.5
11	11.7	11.5
12	9.7	9.6
13	7.4	7.3
14	5.2	5.1
15	3.4	3.4
16	2.1	2.0
17	1.2	1.2
18	0.6	0.6
19	0.3	0.3
20	0.1	0.1
21	†	†

*Based on plan which requires additional 140 only if *some* errors are found in the initial sample of 60.

†Less than 0.1%.

additional 140, if there are some errors in the initial sample of 60) is different than for one sample of 200, one would expect the interval estimate to be different. Indeed it is!

Thus, the confidence interval estimate of the upper limit at the one-sided 95% confidence level for a single sample of 200 if one error is found in the sample is 2.3%. However, if the two-step sampling plan

Table 2-8 One-Step versus Two-Step Sampling* Confidence Limits

(N = 1000)

[Confidence Level = 95% (Two-Sided)]

First Sample n_1	Second Sample n_2	Total Sample n_3	Lower Limit		Upper Limit	
			One-Step	Two-Step	One-Step	Two-Step
			Percent in Sample = 0.5%			
40	100	200	†	0.1	2.5	8.6
60	140	200	†	0.1	2.5	5.8
80	120	200	†	0.1	2.5	4.3
100	100	200	†	0.1	2.5	3.5
			Percent in Sample = 1.0%			
40	160	200	0.1	0.2	3.3	8.6
60	140	200	0.1	0.2	3.3	5.8
80	120	200	0.1	0.2	3.3	4.4
100	100	300	0.1	0.2	3.3	3.7
			Percent in Sample = 2.0%			
40	160	200	0.7	0.7	4.7	8.6
60	140	200	0.7	0.7	4.7	5.9
80	120	200	0.7	0.7	4.7	5.1
100	100	200	0.7	0.7	4.7	4.8
			Percent in Sample 5.0%			
40	160	200	2.7	2.7	8.6	9.4
60	140	200	2.7	2.7	8.6	8.7
80	120	200	2.7	2.6	8.6	8.6
100	100	200	2.7	2.6	8.6	8.6

*Based on an initial sample of size n_1 with an additional sample of n_2 *only* if errors are found in first sample. See Technical Appendix 1 for the computational method.

†Less than 0.1%.

described above is used and one error is found in the sample, the *true* upper confidence limit at the one-sided 95% confidence level is 4.9%, or more than twice that for a single sample. (See Appendix A.)

Thus, an auditor using the two-step plan and finding one error but evaluating with the traditional tables would conclude that most probably the error rate in the population would not be more than 2.3% when it could be up to 4.9%.

This disparity between the assumed limits and the true limits becomes greater as the initial sample size grows smaller compared to the additional sample size even though the overall sample size is constant. For instance for the plan $n_i = 40$ and $n_2 = 160$, the true upper limit (95% one-sided confidence level) is *7.2%* as contrasted with the *2.3%* for a single sample of 200 obtained from the usual table if one error is found.

A comparison of single-sample and two-step sample confidence limits in Table 2-8 illustrates this point and also indicates that as the rate of error in the sample increases for a given n_i and n_2, the disparity between the single- and two-step confidence limits becomes less. (See Technical Appendix 2 for proof.) It further demonstrates that the lower limit changes less than the upper limit for a two-step sample.

TWO-STEP SAMPLING TABLES

The method of computation of the two-step confidence limits is very complex (see Technical Appendix 1). It is not practical for the auditor to routinely attempt to calculate these limits manually. A computer is required to make this process practical.[22] However, a limited set of tables is provided for this purpose.

The availability of a set of tables giving the exact confidence limits for two-step sampling can serve to reduce the sample size requirements and hence the cost of audit tests.

The savings would arise from the fact that in many situations the test indicates acceptable compliance on the basis of a relatively small sample compared with that required if a single sample is used to accomplish the required objective and avoid inconclusive results.

However, there are almost innumerable possible combinations of first and second (incremental) sample sizes as well as criteria for

[22] Akresh and Finley have developed a computer program for two-step sampling evaluation and indicate its availability on the Network Software System of General Electric. See A. D. Akresh and D. R. Finley, "Two-Step Attributes Sampling in Auditing," *The CPA Journal*, December 1979.

extending the sample. A limited working table, providing these two-step confidence limits (interval estimates) when the sample is increased *when more than zero errors are found in the first sample,* is given in Appendix A.

It is to be noted that other types of two-step sampling plans are possible. For instance, instead of requiring zero errors in the first sample to declare the population acceptable, it might be that as many as one (or perhaps two, etc.) could still result in a decision that the population is acceptable.[23] Of course such plans would have different characteristics. It is not practical to attempt to provide tables for all of these plans. They would be too large.

Here is an illustration to demonstrate the use of these tables.

An auditor wishes to test compliance with the system which protects the integrity of payments to vendors in relation to the validity of accounts payable. The auditor decides to rely only partly on the system and will therefore be satisfied with a confidence level of 90%. (See page 35 for a discussion of the selection of confidence levels for compliance test.) The sampling units will be invoices payable which will be examined with respect to a variety of aspects such as proper authorization for the purchase, evidence of receipt of the materials covered by the invoice, proper pricing, etc. There are 10,000 invoices.

For the purposes of this example an MTER of 3% is set for all of these characteristics. (Ordinarily the MTER will vary with the type of error.)

The two-step sampling method is to be used. By examining a table of upper confidence limits for a zero rate of occurrence of error (see Table 2-9), the auditor selects an initial sample of 80 units. If no errors are found in the sample, the upper limit at the 90% confidence level will be 2.8% and thus provide assurance that the rate is less than the MTER (3%).

Assume that as the result of the examination of 80 randomly selected sampling units the following number of errors are found for each characteristic examined.

Type of Compliance Error	Number of Occurrences Found in Initial Sample
A	0
B	4
C	2
D	3

[23]However, the plan with zero in the sample indicating acceptability will result in the lowest sample size when acceptance can be achieved on the initial sample.

Table 2-9 ***Upper Confidence Limits, Single Sample, with Zero Occurrences in Sample, at Various Confidence Levels (%)***

Sample Size*	Upper Confidence Limit		
	90%	95%	99%
	N = 1000		
50	4.4%	5.7%	8.6%
60	3.7	4.7	7.2
70	3.1	4.0	6.1
80	2.7	3.5	5.4
90	2.4	3.1	4.8
100	2.2	2.8	4.3
	N = 10,000		
50	4.5%	5.8%	8.8%
60	3.8	4.9	7.4
70	3.2	4.2	6.3
80	2.8	3.7	5.6
90	2.5	3.3	5.0
100	2.3	2.9	4.5

*Since the sample sizes in the above table are very small compared to the population size for population of 10,000 or more, the population size has a very limited effect on the upper limits. Therefore, the section of the table for N = 10,000 can be used for populations of 10,000 *and over*.

Since no errors were found in the initial sample for error type A, the population is deemed acceptable since there is a 90% probability that the rate of occurrence in the population is less than 2.8% or definitely below the MTER of 3%. *No further sampling is required.*

The results for error types B, C, and D do not permit acceptance but do *not* provide evidence that the error rate for each of these types of errors in the population is higher than the MTER.

If these results are now evaluated as a single sample of 80 units at the 90% confidence level, the resulting interval estimates are shown below.

Compliance Error Type	Number of Errors in Sample of 80	Confidence Interval (%) at 90% Confidence Level
B	4	1.7%-12.0%
C	2	0.4 - 7.6
D	3	1.0 - 9.4

It is evident that the sample of 80 units alone does *not* provide a useful conclusion.

Using the two-step sampling approach, the auditor decides to take an additional sample of 120 units for error types B, C, and D to attempt to obtain conclusive results. The results of this additional audit examination are shown below.

	Number of Occurrences		
Error Type	Initial Sample (80)	Incremental Sample (120)	Total Sample (200)
A	0		
B	4	7	11
C	2	2	4
D	3	6	9

Reference to Appendix A, Table A-2, page 185, provides the following confidence intervals for a two-step sample with an initial sample of 80 and an *additional* sample of 120 at the 90% confidence level for each of the number of occurrences shown in the Total Sample column.

			Confidence Intervals at 90% Confidence Level	
Error Type	Number of Occurrences in Both Samples	MTER, %	Lower Limit, %	Upper Limit, %
B	11	3	3.1	8.9
C	4	3	0.7	4.7
D	9	3	2.4	7.7

The confidence interval estimate for error type B clearly indicates that there is a high probability that the error rate in the population is at least as high as the MTER and most likely higher. This condition is clearly unsatisfactory.

The confidence interval estimate for error type C still provides no conclusive results as to whether the population rate of occurrence is above or below the MTER. It must be emphasized that these tables may not be used to establish confidence interval estimates for a still larger sample.

The confidence interval estimate for error type D, 2.4 to 7.7%, while not providing conclusive evidence that the occurrence rate in the population is above the MTER, does provide an indication that there is a high probability that the error rate is at least marginal (at least 2.4% compared to an MTER of 3%).

COMPLIANCE TESTS AND THE CONFIDENCE LEVEL

The requirement of deciding on a confidence level[24] has been one of the troublesome areas in applying statistical sampling in audit situations.

The confidence level can loosely be described as the probability that the rate of occurrence in the population will be included within the projected confidence interval. More strictly the statisticians would say that with repeated sampling the confidence level will indicate the percentage of the samples which would produce interval estimates containing the population parameter.

Since a statistical sample cannot produce a confidence interval estimate without the specification of a confidence level, such a level must be selected.

As long as the confidence level in a compliance test is reasonably high, when the confidence interval estimate is used in the manner previously described, adequate conclusions may be reached. In the audit situation confidence levels for compliance tests are usually 90% or higher although some have suggested that, in certain situations, an 80% confidence level may suffice.

One criterion that might be examined relates to the use of sample projections in legal cases. It must be remembered that in a civil suit all that is required is a preponderance of evidence, not certainty. Certainty cannot be attained for sample data but probabilities of 90% or more certainly constitute a preponderance of evidence.

It is evident that the confidence level selected should be related to the importance of the compliance test characteristic being tested. In *SAS 1*, it is stated that "the relative weight to be given to the respective sources of reliance and accordingly, the sampling reliability desired for [the] test of details are matters for the auditor's judgment in the circumstances."[25] It is further stated that "the reliability levels used in sampling applications in other fields are not necessarily relevant in determining appropriate levels for applications in auditing because the auditor's reliance on sampling is augmented by other sources of reliance that may not be available in other fields."

There are several basic approaches to the selection of the confidence level for a compliance test. It must be remembered that a compliance test is preceded by a qualitative evaluation of the internal control system to establish its ability to prevent material error *when it is fully and*

[24]The AICPA publications use the term "reliability level" rather than "confidence level."

[25]*Statement on Auditing Standards 1*, p. 39.

correctly operative. This evaluation is then followed by a test to establish that it is indeed so operative.

There are those who would hold that whether the system is fully operative as prescribed is a factual question which is independent of the quality of the internal control system and the protection it provides. Hence, the confidence level should be selected independently from the evaluation of the quality of the system.

One large accounting firm has accepted this philosophy and performs all tests of internal control using a 95% confidence level. This is unquestionably a sound approach since the protective quality of an internal control system is effective only if the system is followed completely and without a serious frequency of deviation.

Others argue that the test of compliance is intended to provide the auditor with information on the extent to which he can rely on the internal control system and hence the extent to which he can limit the scope of his audit examination.

If this second view is accepted the confidence level should vary with the qualitative evaluation of the usefulness of the internal control system or the extent to which the auditor intends to rely on it.

Thus, one possibility would be to use confidence levels in the range of 95 to 99% when the internal control system is rated as excellent and to be relied on heavily. If the control system is rated as only fair in the protection against material error it provides and is to be relied on only moderately, a confidence level in the 90 to 95% range would be used. If the internal control system is rated as poor and not to be relied on, no test at all would be performed.

A philosophy directly opposite to the above also has been proposed. When this approach is used, the confidence level varies *inversely* with the qualitative rating of the system.

Here it is argued that when *prima facie* evidence is obtained from other sources of information such as "walk-throughs," from discussions with employees, and other investigations, that the internal control system is good, a greater risk can be accepted and hence a *lower* confidence level will serve.

On the other hand, it is suggested that when there is a weakness in the internal control system, the testing should be more rigorous and a higher confidence level should be used.

The difference in the last two approaches really lies in a difference in view as to the function of a compliance test and in part to the concept that the compliance test does not stand alone in its evaluation of the operation of the internal control system.

If the view is accepted that the compliance test is an objective examination testing the operation of the internal control system rather than its quality, the first two philosophies, a fixed confidence level for all tests or a confidence level which varies directly with the qualitative rating of the system, would be considered appropriate.

When the view is taken that a weak system requires more testing to provide the protection necessary, the last approach would be acceptable. However, this approach does not really recognize the compliance test as a test of the operation of the internal control system. It views it as further protection against weaknesses in the system. In consideration of the AICPA pronouncements as to the need for testing the *operation* of the internal control system, this last philosophy seems the least defensible.

Nevertheless, the auditor will have to decide as a matter of policy which of these approaches to use in selecting a confidence level.

INTERIM COMPLIANCE TESTS

It is not uncommon for compliance tests by independent public accountants to be performed at some interim point during the audit period, prior to the date of the financial statements about which an opinion is to be rendered.

It is obvious that an interim test performed at a date prior to the end of the audit period must be based on sampling the records for a portion of the audit period with no coverage of the period following the interim date.

Yet there is a basic and generally accepted statistical principle that a sample may be projected only to that portion of the population actually sampled. Thus, an interim compliance test performed statistically, or on a judgment basis, will provide information about adherence to internal controls only for the period sampled. No conclusion can be drawn about the period subsequent to the interim date. To draw conclusions about compliance for the entire audit period, the records for the *entire* audit period must be sampled.

It is noted in *SAS 1* that: "As to accounting control procedures that leave an audit trail of documentary evidence of compliance, tests of compliance . . . ideally should be applied to transactions executed throughout the period under audit because of the general sampling concept that the items to be examined should be selected from the entire set of data to which the resulting conclusions are to be applied."[26]

However, *SAS 1* then states:

> Independent auditors often make such tests during interim work. When this has been done, applications of such tests throughout the remaining

[26]Ibid., p. 30.

> period may not be necessary. Factors to be considered in this respect include (a) the results of the tests during the interim period, (b) responses to inquiries concerning the remaining period, (c) the length of the remaining period, (d) the nature and amount of the transactions or balances involved, (e) evidence of compliance within the remaining period that may be obtained from substantive tests performed by the independent auditor or from tests performed by internal auditors, and (f) other matters the auditor considers relevant in the circumstances.[27]

In other words, while subscribing to the concept that a compliance test "ideally" should cover the entire period, the suggestion is made that sampling of the period after the interim test can be avoided by using alternative audit considerations.

This view is contradictory. It leads to the conclusion that if such alternative considerations can eliminate the compliance test of the later portion of the audit period, then the same considerations could lead to the elimination of compliance sampling for the *entire* audit period. These considerations could replace compliance testing when the previous year's statements are used as the starting point.

If failures to comply with or deviations from the system should occur, or if errors are made, they must begin at some point in time which could just as well be during the period after the interim test.

In fact, there may be an increased probability of failure to comply during the period just prior to the year's end when it is likely to be more hectic and, if the calendar year is the audit period, there is likely to be a greatly increased volume of business.

If, indeed, there is an important failure in compliance with internal controls resulting in a material error, or if there is fraud during the period subsequent to the interim date, the auditor who failed to test this period would be in an indefensible position.

There is no substitute for sampling the entire period, if a conclusion is to be drawn about the entire period, regardless of the problem of sampling the later part during the final audit work period.

If a statistical sample is used there are several alternatives for handling this problem.

Of course, an entirely new sample could be taken from the period following the interim date. However, since the sample requirement to draw a conclusion for that period alone would be about the same as that for the prior period this would be rendered impractical by the cost of the sampling.

Another approach would be to select a sample for the entire audit period but to audit the sampling units included in the period prior to

[27] Ibid.

the interim date at the interim date and to audit the smaller number of records subsequent to the end of the audit period. This method requires that an approximation of the total population size for the entire audit period be made at the interim date. Random numbers not extending beyond the estimate are then drawn for sample selection at that time.

Since the eventual population size is not actually known at the interim date, it will be necessary to reasonably overestimate the population size for this purpose. Thus, at the time of the examination of the sampling units for this later period, it will be found that some sampling units may not exist. Hence, the sample will be a little smaller than that desired. This loss can be anticipated by slightly increasing the desired sample size. A projection can then be made for the entire audit period.

Another approach is to deal with the samples from each period as a stratified sample. Independent unrestricted random samples are drawn from each portion of the population and combined, using the techniques of stratified random sampling, with the two periods constituting separate strata. It is to be emphasized that this method does *not* require a proportionate sample from each stratum but the sample for the second period should be sufficient to provide a reasonable probability of finding instances of failures of the system if they exist.

COMPLIANCE TESTS AND THE GOVERNMENTAL AUDITOR

Generally, the governmental auditor in the United States is *not* called upon to express an opinion on the fairness of the financial statements of the audited governmental agency, although this auditor has an obvious interest in the integrity of the reported book record value.

Governmental auditors are frequently more interested in the operational audit aspects. Thus, their interest is somewhat different from that of independent public accountants.

If welfare payments are being audited, governmental auditors may be interested in the frequency with which ineligibles receive such payments. They may be interested in establishing the frequency with which there has been a failure to fill requisitions for army units on time.

Their determinations are based solely on the sample results. They will no doubt audit other aspects of the activities of the agency under examination, but, unlike public accountants, must draw conclusions about various activities separately.

The use of sample projections will be different for the governmental auditor.

Unlike the independent public accountant who renders a certificate indicating his opinion as to the reasonableness of the financial statement, the governmental auditor issues an audit report calling attention to excessive errors, failures to follow regulations, or lack of compliance with the law. These audit reports highlight important failures and may suggest remedial action.

Thus, the public accountant uses sampling primarily to provide comfort and protection for his expressed opinion on the reasonableness of the financial statement, and the sample results are primarily for his own use.

The governmental auditor must provide specific facts as to the extent of the failure and will usually incorporate the sample results in his audit report to substantiate his audit finding. It is not merely an accept-reject decision.

Governmental auditors must provide in their audit reports some indication of the level of error or failures found. Thus, an estimate of the frequency of such deviations must be established. When based on a statistical sample an interval estimate is required. Such an estimate must be unbiased, objective, and defensible.

Like their colleagues in the commercial field, governmental auditors are usually confronted with a lack of time and resources allocated for their tests. Thus the sample sizes that can be accomplished usually result in wide intervals because of their estimates.

Since they cannot restrict this information to themselves but must publish the results in an audit report to be distributed to a number of organizations, including that audited, some method of representing the facts usefully must be available.

This can be done by using the interval estimate on a *one-sided* basis. Thus, if an unacceptable situation exists they can use the *lower* limit of the estimate alone as an indication that the rate of deviation is *at least* the specified rate. In reporting on an acceptable situation the upper limit of the interval estimate can be stated alone as the *maximum* rate that exists in the population. Of course, in each instance the one-sided confidence level must accompany the statement.

Chapter 3

Substantive (Dollar Value) Tests

The "substantive test" is defined in *SAS 1* as consisting of "two general classes of auditing procedures: (a) tests of details of transactions and balances and (b) analytical review of significant ratios and trends and resulting investigation of unusual fluctuations and questionable items."[1]

According to that publication, these tests, at least in part, provide the evidential matter required by the third standard of field work which specifies that "sufficient competent evidential matter is to be obtained."

It is the first of these methods (a) which comprises the substantive tests referred to in this book. These tests relate to the values the auditor encounters and therefore usually are dollar value estimates. These tests are sometimes referred to as tests of account balances, although they comprise a wider variety of applications.

Since these substantive tests are usually based on the partial examination of records and transactions, the test is a sampling process resulting in the projection of an estimate, usually, of a total dollar value. The technique to be used is estimation sampling for variables.

The projection of dollar values from statistical samples in the audit situation can serve several purposes:

1. To establish the reasonableness of a stated book record value.
2. To estimate an unknown value for use in a financial statement where a book record value is not available or where such a value is known to be unreliable.
3. To establish an amount due, where such an amount is to be collected from some other organization or individual on the basis of a sample projection.

[1] *Statement on Auditing Standards 1*, p. 34.

The third alternative occurs largely in government audits in establishing disallowances in contractor's stated costs, taxpayer's deductions, etc.

Independent public accountants will use dollar value sampling (estimation sampling for variables) differently from governmental auditors.

SUBSTANTIVE TESTS BY INDEPENDENT PUBLIC ACCOUNTANTS

SAS 1 says that "the second standard does not contemplate that the auditor will place complete reliance on internal control to the exclusion of other auditing procedures with respect to material amounts in the financial statements" and that "regardless of the extent of reliance on internal accounting control the auditor's reliance on substantive tests may be derived from tests of details. . . ."[2]

To meet this requirement, the variables (dollar value) sampling procedure may be used to establish the total dollar value of the error or the true (audited) total value by sampling the records or transactions from which the book value was generated.

For instance, when it has been observed that there are errors in counting or pricing in a physical inventory which resulted in an inventory value used in a financial statement, estimation sampling for variables can be used to establish the dollar value of the errors or the true (audited) inventory value to determine whether these failures result in a material error.

In addition, the variables estimation sample can be used for such activities as aging accounts receivable, establishing reserves, etc., in lieu of the costly and often impractical 100% examination.

Beyond the direct application to the audit determinations, the auditor may well encounter values on financial statements which themselves had been projected from statistical samples taken for intercompany settlements or similar purposes.

The auditor may become involved in tax aspects of the use of statistical sampling, where tax return data are created by such samples, as allowed by the Internal Revenue Service, or where the business tax return is audited by the tax auditor using statistical sampling.

Variables sampling has been used for a wide variety of objectives in audit tests and in establishing accounting values. A *few* of these applications are given below as an *illustration* of the kinds of applications which have occurred:

[2] Ibid., pp. 34–35.

1. Inventory observations and evaluation
2. Confirmation of accounts receivable
3. Reserves for:
   a. Bad debts
   b. Inventory obsolescence
   c. Unexpired subscriptions
   d. Unredeemed trading stamps
4. Goods-in-process inventory value
5. Value of fixed assets in a public utility company
6. Aging accounts receivable
7. Intercompany settlements
   a. Airline ticket sales
   b. Rental charges for joint use of telephone and power poles by utilities and telephone companies
8. Dollar value of payments to ineligible welfare and food stamp recipients
9. Disallowances of contractors' costs and expenses by government audits
10. Disallowance of deductions for business tax return data by tax auditors
11. Determination of current portion of installment sales for revolving credit funds
12. Average lines of miles used in local and toll calls for revenue allocation between independent telephone companies and the "long lines" system
13. Inventory of board feet of lumber in forest lands
14. Determination of sales of certain types of merchandise to certain classes of charge account customers over a long period of time for a class action legal case
15. Determining current unearned revenue from ticket sales by airlines

Some of these applications will be discussed in detail later in this volume.

THE DOLLAR VALUE PROJECTION

A total dollar value is projected from the sample, in the direct projection method, by obtaining the point estimate of the total through multiplying the average (arithmetic mean) of the audited values for the sampling units in the sample by the population size. If the direct projection method is not used, the average value of the errors will be determined and multiplied by the population size.[3] The sampling error of the total is then computed and added to and subtracted from the point estimate of the total to provide an interval estimate of the audited value or total amount of error of the population of records or transactions.[4]

The point estimate (the single total value projected from the sample) again is of limited value unless the sampling error is innocuous, an unlikely circumstance in most audit situations. However, the interval estimate limits can often be used in varying fashions depending on the audit objective.

VERIFICATION OF BOOK VALUES

When a substantive test is used by an independent public accountant to establish the reasonableness of a book record value, a very precise projection is not required.

Assume that the book record value of an inventory is stated as $2,482,629.10. To be able to say that this is a reasonable figure, it is not necessary to know the true (audited) value to within the nearest cent nor for that matter the nearest $10 or $100, etc. However, if the estimate of the audited inventory value is known only to the nearest million dollars, the projection is of no value in establishing the reasonableness of the book record value.

To be useful the sampling error must be such that the confidence limits, with a probability equal to the confidence level, would make possible the detection of a material error. The term "materiality" is used here in the accounting sense.

It should be remembered that the amount which is considered to be a material error remains an unresolved problem in accounting. At present, the determination of this amount is largely based on the exercise of judgment by the auditor. Hence the amount fixed for this purpose in audit tests should be considered not only to be approximate but indeed a general area rather than a precise figure.

It is desirable that the permissible sampling error (sample precision), in order to provide useful results for this purpose, be equal to one-half

[3]This essentially is the same as a difference estimate. Similar results are obtained by the ratio estimate approach. See H. Arkin, *Handbook of Sampling for Auditing and Accounting*, 2d. ed., McGraw-Hill Book Company, New York, 1974, chap. 12.

[4]Ibid., chap. 7, or any standard statistical text will provide information on the method of computing sampling errors.

of the smallest amount considered to be material by the auditor, for reasons explained later.

It may well develop that the size of the sample required for this purpose is not practical or within the capability of the auditor in a given situation. In that event, a larger sampling error may be permitted but the risk of failing to detect a material error is increased. The risks involved are discussed in detail in Chapter 8.

If the book record value is found to be within the interval estimate projected from the sample, there is no basis for declaring that it is either overstated or understated. If a sampling error equal to one-half of the smallest amount considered to be material has been used, and the book record value is within the confidence interval, the book record value will never be further away from either confidence limit by more than the amount considered to be material. This will provide adequate protection against a material error.

The various statistical and audit risks involved in this type of test will be discussed in detail later. (See Chapter 8.)

INTERPRETING THE RESULTS

Assuming the sample error is not more than half the material amount, the following conclusions can be drawn by a comparison of the confidence interval with the stated book record value as shown in circumstances indicated below:

1. Book record value is *within* the confidence interval (between the upper and lower limits).

   Conclusion: There is no basis for concluding that the book record value is misstated. However, if the distance between the book record value and the *farthest* confidence limit is an amount considered to be material, no audit conclusion can be drawn.

   An audit conclusion that a book record value is fairly stated can be obtained *only* when the book value falls between the upper and lower limits *and* the difference between the book record value and the furthest limit is *less* than the smallest amount considered to be material.
2. Book record value is *above* the upper confidence limit.

   Conclusion: Most probably (with probability at least equivalent to the confidence level) the book record value is overstated by *at least* the difference between the *upper* limit and the book record value.
3. Book record value is *below* the lower confidence limit.

Conclusion: Most probably (with a probability at least equal to the confidence level), the book record value is understated by at *least* the difference between the *lower* limit and the book record value.

VARIABLES SAMPLING AND RARE VALUES

When using variables sampling to establish the reasonableness of a book value, special care must be exercised to assure that adequate protection is provided against the situation where there might be one or a very few sampling units among a great many which in total constitute a material error.

Obviously, if none of these errors are included in the sample, the sample projection will not reflect the error in the population. Yet the probability that rare but material errors will be included in a small sample from a large population might be very low. For instance, if there is one transaction among 10,000 which constitutes a material error, the probability that it will be included in a sample of 100 is only 1%. If there are 10 errors totaling a material amount among the 10,000 entries, the probability that any will be included in the sample of 100 and thus be projected is only 9.6%.

To provide some protection against this situation, it is highly desirable that all items which could in themselves include a material error or one close to that amount be examined on a 100% basis, with other items sampled. This procedure is usually followed by auditors anyway on the grounds of materiality. It is wise to incorporate this approach in *all* variable sampling plans.

In addition, this danger emphasizes the need for samples of reasonable size regardless of the outcome of calculations for sample sizes, so that infrequent errors may be included in the sample. It is well in a variables sample, apart from the usual method of sample size determination, to consider the sample size from a discovery sampling viewpoint.

Assume a situation in which the records to be audited to establish reasonableness for the book value are a group of vouchers.

The total book record value of these vouchers is $2,156,218. There are 3019 vouchers in the defined population. The auditor decides that the minimum amount of error which would be considered to be material is $125,000. It is known that there are four vouchers which individually amount to more than $125,000. It is evident that these large vouchers should be dealt with separately on the basis of a 100% examination.

In addition, among the remaining 3015 vouchers there are approximately 125 vouchers with values of over $10,000. Thus, if even 30 of these vouchers were overstated by as much as 50% of their value, a material error would exist. It would be essential that at least one of these 30 be included in the sample.

It is assumed that no stratification, other than the identification of the four largest accounts, is feasible in this situation. Therefore, it is decided that an unrestricted random sample will be used for the remaining 3015 vouchers.

The 30 accounts make up about 1% of the 3015. Resort to a discovery sampling table indicates that to achieve a 90% probability of including at least one of these possibly overstated vouchers in a population of about 3000, a sample of 222 would be required.[5]

Of course, other criteria could be used in establishing this minimum sample size. For instance, a more conservative approach could be based on the assumption that 15 of these 125 vouchers might be overstated by as much as $10,000 each, resulting in an error material in amount. The 15 such overstated accounts make up 0.5% of the sampled population. The discovery sample table indicates that the sample size required to provide a 90% probability of including at least one of these in the sample is 422.

The auditor, using this type of reasoning, can establish a *minimum* requirement for a sample size to protect against the possibility of not detecting a material error when a very few items might constitute such an error collectively.

Of course, the overall sample size estimated to be required for this test would be determined in the usual way, but the *minimum* requirement regardless of that result would be established by the above reasoning.

If stratification is feasible, especially when more than two strata are used, a disproportionately large sample could be drawn from the stratum containing the 125 vouchers to overcome this difficulty. The sample size for that stratum might then be sufficient to assure such disclosure.

ADJUSTMENT OF BOOK VALUES

If a statistical sample produces a confidence interval which does not include between the projected confidence limits the book record value, there is a high probability that the book record value is misstated. If a sufficiently high confidence level has been used, the auditor can then consider this book value to be incorrect.

At this point, the question of a possible adjustment to the book record value by the client may be raised. If this adjustment is to be under-

[5]Such a table can be found in H. Arkin, op. cit., p. 470.

taken, an adjustment amount must be established which objectively and defensibly can properly be applied to the book record value.

If the spread of the confidence interval is inconsequential (very small sampling error), the point estimate of the total projected from the sample may be used as the corrected book value.

Confronted with the usual audit limitations of time and cost, the interval estimate of the audited value will probably involve a considerable spread between the upper and lower limits of the confidence interval estimate.

A small sampling error is not likely to be encountered in the audit situation. If the spread of the interval is of consequence, even though less than a material amount, the use of the point estimate may not be defensible. While technically the point estimate is the most likely single value, no specific probability can be assigned to that value and it certainly would be a very small probability as compared with the wide range of other values within the limits.

While it may be desirable to consider extending the sample to achieve a negligible or very limited sampling error, this in many, if not most, cases may not be feasible because of time and cost considerations. In addition, if an inventory book value has been established to be overstated by sample projection, the inventory date may have passed and the inventory changed before this fact is known.

If it is necessary to establish the amount of adjustment from a projection of an audit sample, and there is an appreciable sampling error, such that the spread is significant, the appropriate action is to adjust the book record value to the confidence limit nearest the stated book record value.

Assume a situation where a book value is given as $2,647,264.10. An estimate from a statistical sample is projected to provide an interval estimate of the true (audited) value to be between $2,436,375 and $2,536,375 at the 95% confidence level. The book record value is clearly overstated. It is assumed that, as usual, the sample size is restricted by limitations of time and resources. In this case, the adjusted book record value would be established as $2,536,375 (upper limit).

The principle involved is that the projection from the sample indicates that there is a high probability that the book record value *at most* is equal to $2,536,375. While it is very probable that the actual value is less, there is no way of establishing this lower value with any high probability of exactitude unless an impractically large sample is used. Thus, an adjustment to any lower value *might* introduce an error in the opposite direction. An adjustment to a lower value would not be defensible legally.

If, as previously suggested (see page 45), the sampling error is established to be one-half of the smallest amount considered to be material, the probability that there will remain a material overstatement in the book record value is very small. Of course, if a larger sampling error is permitted, there is an increased probability that a material overstatement remains. However, if the sampling error is reasonable, though more than one-half of the material amount, this risk may not be great. A discussion of these risks was given earlier (see page 44).

If the book record value had been $2,390,150, then a conclusion that it had been understated, and an adjustment to $2,436,375 (lower limit), would have been appropriate.

When the book record value is not included within the confidence interval there is a high probability of a misstatement. In making an adjustment it must be remembered that the nature of the process which establishes the smallest amount deemed to be material is judgmental and by no means exact.

If the book record value is not included in the interval estimate, it would be concluded that the difference between the book value and the sample estimate is *statistically significant*. The fact that the difference is *statistically significant* merely means that there is a very low probability that another sample of the same size would produce a confidence interval which would include the book value and fail to establish the validity of the difference.

However, the concept of *statistical* significance should not be confused with significance in the ordinary sense, where it is commonly interpreted to mean "important" or "material." The fact that the book record value is not included in the confidence interval does indicate a high probability that it is wrong but does not in itself indicate that it is materially wrong.

Whether or not a difference between the book value and the sample projection is material will depend on the magnitude of the sampling error or spread of the confidence interval. If the spread between the upper and lower limits of the confidence interval estimate is *less* than the smallest amount considered to be material, the difference may be statistically significant, but not material in the accounting sense.

The only requirement needed to establish that *any* difference, no matter how small, is *statistically* significant is a large enough sample. On the other hand, the difference cannot be considered material, no matter how large, unless it is proven to be statistically significant.

To merit an adjustment, the difference must be demonstrated not only to be statistically significant but material as well.

If, and only if, the difference between the book value and the *furthest* limit of the confidence interval estimate is equal to or larger than the smallest amount considered to be material is there any appreciable probability that a material error *might* exist. An adjustment to a book record value when this condition does not exist, regardless of whether the book value is within the confidence interval, is not appropriate.

CONFIDENCE LEVELS IN SUBSTANTIVE TESTS

A confidence level must be selected in the performance of a substantive test. *SAS 1* states that "for statistical samples designed to test the validity or bona fides of accounting data and to be evaluated in monetary terms, the committee believes the foregoing concept should be applied by specifying reliability levels that vary inversely with the subjective reliance assigned to internal control and to any other auditing procedures or conditions relating to the particular matters to be tested by such samples."[6]

This statement can generally be interpreted to mean that when other audit assurance is available, a greater risk can be taken with respect to the substantive tests.

Of course, as in the case of the compliance test, confidence levels of 90% or higher will be generally acceptable. However, the philosophy suggested above has led to ridiculously low confidence levels in some cases.

The substantive test, although part of a set of complex audit examinations and procedures, is designed to provide an *independent* factual basis for establishing whether a book record value is misstated. As such, the logic of widely varying the risks in such a test is questionable.

Of course, if the internal control system is evaluated as being excellently protective against a material error *and* the test of compliance indicates satisfactory operation of that system, the auditor may decide not to test certain book record values at all. However, if the test is performed at all, it should in itself provide sufficiently low risks of incorrect conclusions with respect to the book value being tested.

On the other hand, some argument may be made that within a suitably high range (say 90 to 99%) some variation in risk may be permissible in terms of the importance of the test to the overall audit conclusion, resulting in the certification or an increased confidence level where other audit evidence makes a specific book record value suspicious.

Nevertheless, the approach suggested in *SAS 1* seems to be that the confidence level from one or more statistical samples should be com-

[6] *Statement on Auditing Standards 1*, p. 40. It should be remembered that the AICPA uses the term "reliability" rather than "confidence" level.

bined with the *subjective* reliance assigned to other audit procedures to establish the overall risk.

It is suggested in *SAS 1* that if little if any reliance is assigned to internal controls for the purpose of a specific substantive test, a 95% confidence level is reasonable.

The *SAS 1* method provides a means of combining a *subjective* evaluation, to be expressed quantitatively, of the reliance to be placed on internal accounting control and "other relevant factors" to calculate a confidence level for a substantive test to achieve an overall confidence level apparently for the *whole audit examination*.

A formula is provided[7]

$$S = 1 - \frac{1 - R}{1 - C}$$

where S = confidence (reliability) level for substantive test
R = combined confidence (reliability) level desired
C = reliance assigned to internal accounting control and other relevant factors based on auditor's judgment

If expressed in terms of the *risk* rather than confidence level, the formula becomes:

$$1 - S = \frac{1 - R}{1 - C}$$

and

$$1 - R = (1 - C)(1 - S)$$

or

$$R' = C'S'$$

where R' = overall risk (1 − CL)
C' = compliance risk
S' = substantive test risk

Thus it is stated that the overall risk is equal to the product of the risk in "reliance on internal controls" and the risk in the substantive test (1 − CL).

No basis is provided for the use of this formula in *SAS 1*. However, it seems to be an attempt to provide a basis in probability theory for establishing a confidence level for a substantive test, taking into consideration the reliance to be placed on the internal control system.

There is considerable doubt as to the validity of such a probability basis for the use of this formula. See Technical Appendix 3 for the technical explanation for this conclusion.

[7]Ibid., p. 53.

For example, assume that an auditor draws the subjective conclusion that there is a 90% probability that the internal control system and "other relevant factors" have prevented a material error in a given audit situation. This means that there is a 10% risk that it did not. He wishes an overall 95% assurance that there is no material error or, in other words, is willing to run a 5% risk. The appropriate risk for a substantive test is then

$$S' = \frac{.05}{.10} = .50$$

and the confidence level for the substantive test is

$$1 - S' = 1 - .50 = 50\%$$

A confidence level as low as 50% will often result in ridiculously low sample sizes which, in light of the prior discussion of the probability of including error items in the sample, will not provide protection.

It is not at all clear how this subjective evaluation of the reliance to be placed on the internal control system and "other relevant factors" is to be expressed as a numerical value. To say the least, this is highly subjective and may vary greatly from auditor to auditor, thus at least partially destroying the objectivity of the statistical sampling approach.

The purpose of a substantive test is to provide the facts upon which the auditor bases a decision. This method provides facts (interval estimates) which may have a very low probability of being correct.

Another method which has been used to determine confidence levels for tests of account balances is to arbitrarily relate the confidence level to the subjective evaluation of the protective ability of the internal control system *after* the operation of internal controls has been found to be effective by compliance tests. The confidence levels which might be used in such a system are:

Confirmed Evaluation of Internal Control System	Confidence Level Range for Substantive Test, %
Excellent	90
Fair	95
Weak	99

This method conforms to the AICPA philosophy that the confidence levels in substantive tests may vary inversely with the evaluation of internal controls. Nevertheless, it keeps the confidence levels within a reasonable range.

A defensible approach in a substantive test is to minimize the exercise of judgment by the auditor to provide an interval estimate that has a reasonably high probability of including the audited population value. This can be done by using confidence levels of 90% or higher regardless of the outcome of other tests and examinations if the substantive test is to be performed at all.

CONFIDENCE LEVELS AND THE MAGNITUDE OF POSSIBLE ERROR

As previously discussed, in relation to compliance tests the choice of a confidence level not only specifies the probability that the true audited value of the population is included between the upper and lower confidence interval limits but it also has an impact on the possible magnitude of the error when the true value is not within the interval.

For the usual confidence levels of 90, 95, and 99% there is a small possibility that the sample actually selected will produce an interval estimate which will not include the true population value. For the above confidence levels, these probabilities are 10, 5, and 1% respectively.

Thus, when an unusual sample (one of the 10, 5, or 1%) is drawn so that the population value is not within the confidence limits, the question may be raised as to how far off this estimate may be.

Low confidence levels for a given sample size from a specified population will produce narrower confidence intervals than will higher confidence levels, thus increasing the chance that the population value will not be included in the sampling interval. This is another way of saying that for a given sample size from a specified population, the lower the confidence level, the smaller the sampling error.

To illustrate this point, assume that a sample is to be taken of a size for which the sampling error of the average at the 99.9% confidence level is $33.[8]

Since there is only 1 chance in 1000 that the sample point estimate will deviate from the population average by more than this amount, $33 will be referred to as *maximum sampling error*. It will be assumed that the probability of an actual departure of the sample average from the population average is so small (1 chance in 1000) that the possibility of a greater deviation may be ignored.

The sampling error at various confidence levels other than 99.9% may now be compared to the *maximum sampling error*.

It will be seen that for a given sample size for a sample drawn from

[8]The standard error is $10. The population size is assumed to be very large.

Table 3-1 ***Comparison of Indicated and Maximum Sampling Errors at Various (Two-Sided) Confidence Levels***

Confidence Level Used, %	Indicated Sampling Error	Maximum Sampling Error	Difference Unusual Sample		Probability of Exceeding Stated Sampling Error, %
			Amount	Percent	
50*	±$ 6.75	$33.00	$26.25	389	50
80	± 12.80	33.00	21.80	170	20
90	± 16.50	33.00	16.50	100	10
95	± 19.60	33.00	13.40	68	5

*The inclusion of a 50% confidence level does not indicate that such a confidence level should ever be used. However, *SAS 1* implies such a possibility.

a given population the difference between the sampling error for a given confidence level and the maximum sampling error grows as the confidence levels decline (see Table 3-1).

It is clear that when very low confidence levels, less than 90%, are used there could be deviations of great magnitude. The risks involved in such a situation in conclusions that there are material errors, when in fact there are not, thus increase as the confidence level actually used declines (see Chapter 4).

PROJECTIONS OF UNKNOWN VALUES

As previously noted (page 42), the variables (dollar value) estimation sampling technique is frequently used in the accounting environment to provide estimates of unknown values. Examples of such usages include aging accounts receivables, establishing inventory values without 100% physical inventories, making intercompany settlements, setting the current portion of installment sales, etc.

In general, when the necessary value is not generated by the accounting system (e.g., reserve for unexpired subscriptions) or when the available value is considered to be unreliable or known to be wrong, the statistical sampling approach may be used in lieu of costly, time-consuming 100% detailing.

While variables estimation sampling may be used in such situations, the approach must be somewhat different for the independent public accountant.

In this situation the point estimate of the total dollar value may be appropriate to establish the required value *providing the sampling error is innocuous.*

If the value projected is to be used in a financial statement or for some equally important purpose, a much smaller sampling error than those usually applied in audit test situations is required. Here there is no book record value to which the sample projection can be compared. It would be expected that the permitted sampling error would probably be *less* than the one-half of the smallest amount to be considered material that was suggested previously for audit tests.

For instance, the Revenue Procedure 64-4 of the Internal Revenue Service, which deals with the sample projection of current sales in a revolving credit fund, prescribes a sample precision which at the 95% confidence level amounts to about 4% of the total value of the projected point estimate of the total value when it is to be used on the tax return.[9]

This level of precision may in some cases be unnecessarily tight. The question of the absolute total dollar value involved may be more important than the relative precision. Nevertheless, if the resulting sampling error is not sufficiently great to restrict its use for financial record purposes, it may be used.

It may well happen that the sample size required to achieve such precision will be prohibitive. The alternative of a 100% detailing in such instances may also be out of the question. In such cases, the confidence limits may be used in the manner described in the next section of this chapter dealing with the use of variables estimates in government audits.

Of course if a prohibitive sample size requirement is encountered the use of some of the more sophisticated sampling methods should be considered. For instance, stratification, or differences, or ratio estimates may be used in certain situations to achieve a large reduction in the sample size requirement.

VARIABLES SAMPLING AND GOVERNMENT AUDITS

Since the governmental auditor and the commercial internal auditor usually are not required to issue an opinion about a financial statement, their view of the substantive (dollar value) test is different from that of the independent public accountant.

Nevertheless, the governmental auditor is equally concerned with the integrity of reported values. The difference is that this auditor is

[9]IRS Revenue Procedure 64-4, U.S. Internal Revenue Service. The requirement given in this publication is that the coefficient of variation is required to be not more than 2%. The *relative* precision can be obtained by multiplying the 2% by the appropriate factor for the selected confidence level.

concerned with individual totals of a specific nature and will draw conclusions about each value separately. A governmental auditor will usually include on the audit report any significant disparities in the individual totals resulting from government operations.

These audit findings are frequently concerned with the failures in operations and may be expressed both in terms of frequency of occurrence *and* the dollar values involved. For instance, in an audit of government welfare payments to individuals, the auditor would include in the report the frequency of payments to ineligible individuals as well as the total dollar values involved.

Another function of these audit findings may be to serve as a basis for collection of payments determined to be improper from the responsible organizations or agencies.

A federal agency auditing welfare payments or food stamp issuances may, as the result of the audit test, act to recover improper expenditures resulting from improper payments by a state or county. A government audit of claimed costs or expenses by a contractor who is reimbursed by the government, on a cost-plus basis, may result in recovery from a contractor. An audit of a business tax return may result in disallowances of a taxpayer's claimed deductions.

In the above instances, confronted with numerous records or transactions, the governmental auditor can examine only a portion of those records. In many government audits, the number of records involved may be in the tens or hundreds of thousands or millions.

If a statistical sample is used, a projection of the total dollar value may be used for these purposes. Of course, an interval estimate will result from the sample projection. If the sampling error for the interval estimate is negligible or very small, the point estimate may be used for all of these purposes. However, such a situation is most unlikely with the sampling the auditor can achieve. The auditor must make use of the confidence limits.

The audit finding included in the auditor's report, in the type of application mentioned above, would probably be based on a one-sided limit. For instance, the auditor might state that the total value of errors of some type was at *least* a specified value, where this value is the *lower* confidence limit. On the other hand, it may be appropriate to state that the total value of the errors was *at the most* equal to the figure determined by the *upper* confidence limit.

There are no fixed rules nor even a universally good approach for establishing the desired sample precision for such purposes. Nor can the confidence level be selected on any basis other than the auditor's judgment.

However, the auditor must remember that the smaller the sample size used, the larger the sampling error and, hence, the wider the interval estimates. Since the confidence limits used are determined by the spread of the interval, restrictions on sample sizes may limit the usefulness of the result in an audit report.

Thus, a very low lower limit, where it is desired to emphasize that the total dollar value is at least a certain figure, may arise merely from an inadequate sample size and defeat the very purpose of the audit. A similar result will arise in the use of the upper limit to establish that *at most* the total is a certain figure. The audit may be defeated by too small a sample. Reasonable precision is required to render useful results. The individual auditor must decide on the precision which might be adequate for his purpose and use a sufficient sample to achieve that end.

As an example, assume that an audit test result produces a point estimate of $751,298 for the total value of unnecessary costs that arose because certain procurement procedures were not properly followed. A sample of 50 purchases from a population of 10,000 was taken. It is found that the standard deviation of these errors amounted to $250. The sampling error of the total value at the 95% confidence level (one-sided) computed from these figures is $581,902. The auditor might report then that the total loss was at least $169,396 at the 95% one-sided confidence level—not a very impressive figure compared to the point estimate. If a sample of 200 had been used and the same results were obtained from the sample, it could be claimed that the loss was at least $462,560 (95% one-sided confidence level), and for a similar sample of 400 with the same values, $549,400.[10]

The selection of the confidence level will be determined by the importance of the data produced. The greater the consequence of the finding, the higher the confidence level should be. As before, confidence levels of 90% or more will be adequate.

The second type of application of statistical sampling to establish the total value of disallowances or improper payments has been used increasingly often.

A tax auditor examining a very long list of individual tax credits or a huge number of invoices for excise tax payments has the choice of auditing just a few of the largest items, or projecting the total disallowance on the basis of a statistical sample. The time available for each

[10]The assumption in this discussion is that the sample values did not change for the larger sample. It should be remembered that if a larger sample is indeed drawn, both the point estimate and the standard deviation of that sample will probably be different than that of the smaller sample.

business tax return is obviously so limited that the auditor cannot examine all claimed deductions for a large business nor in many cases all those for smaller businesses.

If only a limited number of items are examined the disallowance is limited to these individual transactions audited, although the presence of some disallowances in the sample is an almost certain indicator that there are others in the remainder of the population.

When a probability sample is projected, the result is an estimate of the total disallowance in the entire population, a value virtually certain to be considerably greater than that resulting from an audit of individually selected items.

In a recent audit, the examination of the 122 highest valued transactions produced a disallowance of about $15,000. However, a statistical sample of 522 projected to provide an adjustment based on the lower confidence limit (one-sided 95% confidence level) was about $2,420,000 or more than 160 times as much.

In a similar manner, projections of disallowances in audits of a contractor's claimed costs, of various payments, etc., will provide an overall estimate for the population considerably larger than that obtainable by examining a limited number of selected items.

The statistical sample used for such a purpose produces an interval estimate. Unless the sampling error is so small as to be inconsequential to both parties, the point estimate cannot be used as a basis for claimed reimbursement. Again, the auditor's available time and resources simply do not often permit a large enough sample to produce a small sampling error. The confidence limits must be used.

Thus, the auditor who is basing a claim on a statistical sampling projection will find that the only defensible approach is to rest that claim on the *lower* limit of the interval estimate.

The auditor can state that there is a high probability (equal to the confidence level) that the actual amount due is *at least* the specified figure. In fact, there is a very high probability that it is more than the lower limit but a higher figure cannot be established with any high probability. The lower limit is the only value that is likely to be accepted by the courts in a legal contest.

It will be remembered that the lower limit is calculated by *subtracting* the sampling error of the total from the point estimate of the total. Thus, the smaller the sample, the larger the sampling error and the less collected!

This gives rise to a cost-benefit relationship of considerable importance. The larger the sample, the greater the amount collected, but also the greater the cost of the audit.

If the cost per sampling unit audited is known, this relationship can

Table 3-2 ***Audit Cost versus Potential Recovery***

Sample Size	Amount of Claim (Lower Limit)	Overall Audit Cost	Net Gain
100	$20,000	$ 2,000	$18,000
200	43,430	4,000	39,430
300	53,810	6,000	47,810
400	60,000	8,000	52,000
500	64,220	10,000	54,220
600	67,340	12,000	55,000
700	69,760	14,000	55,760
800	71,720	16,000	55,720
900	73,330	18,000	55,330

be developed. Assuming that the point estimate and standard deviation derived from a larger sample remain unchanged,[11] the *decrease* in sampling error of the total and, as a result, the increase in the lower limit and amount collected for a large population will be proportional to

$$\frac{\sqrt{n_L}}{\sqrt{n_s}}$$

where n_L = larger sample size
n_s = smaller sample size

On the other hand, the cost of the larger sample will *increase* in the ratio

$$\frac{n_L}{n_s}$$

The decision to use a larger sample then becomes a function of the audit cost per unit versus the increase in the amount of the claim.

Assume that in an audit of a government contract, a statistical sample of 100 cost-items out of a population of 10,000 produced a point estimate projection of disallowed expenses of $100,000. The sampling error of $80,000 resulted in a lower limit of $20,000, which is to be the amount claimed. If the overall average cost of auditing per sampling unit is estimated to be $20, the relationships that are shown in Table 3-2 result.

It is evident that the point of diminishing returns in this situation occurs when the sample size is about 750. After that point, the *net*

[11]The point estimate and standard deviation of a larger sample probably will be different than that of the smaller sample in a nonpredictable way.

recovery (claim, less audit costs) starts to decline. However, the net gain is quite small prior to that point and probably a lower sample size is more appropriate.

Once again, the assumption has been made that the point estimate and standard deviations of all these sizes of samples will be constant. This is not likely. However, an approximation of the break-even point can be obtained by this method.

THE SAMPLING DISTRIBUTION PROBLEM

The computation of the sampling error for the arithmetic mean (or total), and as a result the interval estimate, is based on the concept of the sampling distribution. The sampling distribution consists of the arithmetic means (or totals) of values from all possible samples of a given size drawn from a specified population.

The usual method of computation of the sampling error is based on the assumption that the sampling distribution from which this value is computed is in the form of a normal distribution. It is apparent that to the extent that the sampling distribution actually departs from normality, the resulting computation of the confidence interval will be incorrect and the probability that the actual population value will be included within the confidence interval will not be the same as that stated for the confidence level.

The *central-limit theorem* indicates that regardless of the distribution of the values in the population, the sampling distribution of the arithmetic average (or total) will tend toward normality as the sample size grows. However, the rapidity with which this occurs is dependent on both the sample size used *and* the nature of the distribution of the values being sampled.

Table 3-3

Dollar Amounts	Number of Invoices	Cumulative Dollar Value
$ 0–499	6072	$ 476,794
500–999	347	791,057
1000–1999	221	1,033,883
2000–2999	82	1,230,336
3000–3999	34	1,344,845
4000–4999	25	1,455,624
5000–5999	20	1,567,031
6000–6999	8	1,618,312
7000–7999	8	1,677,009
8000–8999	8	1,742,136
Over 8500	54	9,633,884

Table 3-4

Account Balances	Number of Accounts Receivable	Cumulative Dollar Value
Under $10	73,520	$ 501,454.43
$ 10– 24.99	34,273	1,034,267.61
25– 49.99	14,766	1,553,155.67
50– 99.99	8,822	2,168,621.56
100– 199.99	4,676	2,818,709.76
200– 399.99	3,838	3,929,393.14
400– 799.99	1,982	4,950,499.94
800–1499.99	224	5,164,617.56
1500 and over	24	5,213,261.31

POPULATIONS OF ACCOUNTING VALUES

Very little information is generally available about the population distribution of the values of accounts receivable, invoices, purchase orders, entries, etc. Virtually nothing is known about the nature of the population of errors in accounting records.

Experience in dealing with accounting entries and records would seem to indicate that generally they consist of considerable numbers of small or moderate values usually accompanied by a very limited number of very high values. For instance, the distribution of the book record values of the invoices of a large company are given in Table 3-3.

As sometimes occurs, the few largest items make up a large proportion of the entire population value. In this case, the 54 invoices of greatest amount constitute about 82% of the total value. This situation accounts for the usual practice of the auditor, who out of concern about materiality, usually examines all the largest values as well as sampling the balance of the values.

However, this degree of concentration in the high value items is not invariably the case, as indicated in Table 3-4 in the distribution of accounts receivable balances of a domestic motor carrier.[12]

In this instance, the 248 largest account balances amounted to only 5% of the total of all accounts receivable. The basic distribution consisting of a large number of small- or moderate-value items and a limited number of high-value items is common, often with a fairly large proportion in the high-value group. This type of distribution is known technically as a positively (right) skewed distribution. The degree of skewness can be measured. (See Technical Appendix 4 for an explanation of the method.)

[12]From W. E. Deming and T. N. Grice, "An Efficient Procedure for Audit of Accounts Receivable," *Management Accounting*, March 1970, p. 9.

The distribution of the book record values will be dependent on the type of account and type of industry involved. A much greater uniformity will be found in some cases. For instance, in auditing payments to individuals under a government welfare program, the maximum payment has an upper limit and most individuals tend to receive relatively similar amounts.

Another feature which characterizes the distribution of accounting values is the occurrence of zero balances (as in accounts receivable) and negative values (credits included in debit balances and debits included in credit balances).

While there seems to be no generally available empirical data for the distribution of errors in populations of accounting records, only for those in samples, nevertheless certain general conclusions can be drawn.

It is certain that even in a bad situation the majority of the recorded values are correct. This results in a high concentration of zero values for errors. Since there can be both positive or negative errors, those values may spread out on both sides of the zero values.

The resulting distribution of values of errors will then have a high concentration of values and will be skewed as well. The degree of concentration of these values can be measured. This characteristic of the distribution is referred to as *kurtosis*.

THE SAMPLING DISTRIBUTION FOR ACCOUNTING VALUES

As previously noted, the computation of the sampling error of the arithmetic mean (or total) which is used to establish the confidence limits is based on the assumption that the sampling distribution for these values is a normal distribution. This form of frequency distribution is also known as the *bell-shaped curve*. The sampling distribution consists of the averages (or totals) computed from all possible probability samples of a specified size drawn from a specified population.

According to the central-limit theorem, the sampling distribution of totals estimated from the sample should tend to form a normal distribution as the sample size grows regardless of the nature of the population. However, since this occurs only as the sample grows, for small samples the sampling distribution will tend to be similar to the distribution of the sampled population but less skewed. It will tend to be right skewed for accounting data except for rare kinds.

The sampling distribution of arithmetic averages (or totals) of *errors* in accounting documents will tend to be highly concentrated (peaked) at zero and may be skewed as well for small samples. If the point esti-

Table 3-5

Dollar Amounts	Number of Stock Items	Cumulative Dollar Value
$ 0– 99	316	$ 33,245
200– 499	219	110,068
500– 999	150	219,459
1000–1999	90	344,782
2000–4999	61	532,350
5000–9999	30	733,975
10,000 and over	26	1,302,720
	892	

NOTE: This distribution does not include a group of 174 subassemblies.

mate of the average (or total) is a positive figure, the sampling distribution generated by sampling a population of errors in accounting records will be right skewed as well.

However, the question may be raised as to how large the sample must be to achieve a sufficiently close approximation to the normal curve to make the calculated confidence limits accurate enough for practical purposes in accounting operations.

Populations of accounting values are almost always positively (right) skewed and frequently drastically so. This skewness can be measured by a value specified as α_3. (See Technical Appendix 4 for its formula and the method of computation.) The value of α_3 is equal to zero for normal distribution, but may run as high as 6 or 7 in a very badly skewed population of accounting values (invoices, accounts payable, etc.).

For the distribution of values of invoices specified in Table 3-3 the value of α_3 is 6.3, when the highest 54 values are removed for separate audit. For the population of accounts receivable shown in Table 3-4 the α_3 value is 5.2 when the top 24 values are handled separately. However, these distributions will be found to be more skewed than many distributions of accounting values.

For instance, a distribution of inventory values is shown in Table 3-5.[13] When the 26 highest-value items ($10,000 and over) are audited on a 100% basis, the measure of skewness (α_3) of the remaining distribution is 3.1.

[13]W. E. Courtright, and A. A. Procassine, "Inventory Valuation by Sampling," *Industrial Quality Control*, February 1958, p. 18.

Little is available on the measurement of skewness of accounting errors, but some general conclusions can be reached about their distribution. If less than one-half of the sampling units are in error and with a positive point estimate, the skewness will be positive.

As previously noted the errors can be positive or negative. Those sampling units which have no errors are assigned a value of zero. In the discussion below it is assumed that the sign of the individual errors (+ or −) will be assigned so that the point estimate is always positive.

In a recent audit of errors in payments to individuals by a government agency the value of skewness (α_3) was 3.6.

The skewness of a sampling distribution which results in a distortion of the actual confidence limits compared with those obtained from the normal assumption is known to be:[14]

$$\alpha_{3\underline{\bar{X}}} = \frac{\alpha_3}{\sqrt{n}}$$

where $\alpha_{3\underline{\bar{X}}}$ = skewness of sampling distribution of arithmetic averages
α_3 = skewness of population values
n = sample size

Thus, the skewness of the sampling distribution declines as the sample size grows.

For the two previously mentioned distributions of accounting values (invoices and receivables), for samples of 100, the sampling distributions would have a value of α_3 (measure of skewness) of .63 and .52 respectively. For the inventory values in Table 3-5 the skewness of the sampling distribution of averages of samples of 100 would be only .27. For the errors in the government audit, the skewness (α_3) of the sampling distribution would be .36. This would indicate a moderate degree of skewness for the sampling distribution as compared to zero for a normal distribution. For samples of 400 the α_3 values for the sampling distribution would be only .32, .26, and .14, respectively.

If the sampling distribution is positively (right) skewed, the use of the normal approximation to establish the confidence limits will result in an understatement of *both* the upper and lower confidence limits. While the exact sampling distribution drawn from a skewed population is not known, it may be estimated. (See Technical Appendix 5 for an explanation of the method.) An example is shown in Figure 3-1. It is

[14]For a more technical discussion of this formula see I. W. Burr, *Engineering Statistics and Quality Control*, McGraw-Hill Book Company, New York, 1953, p. 171.

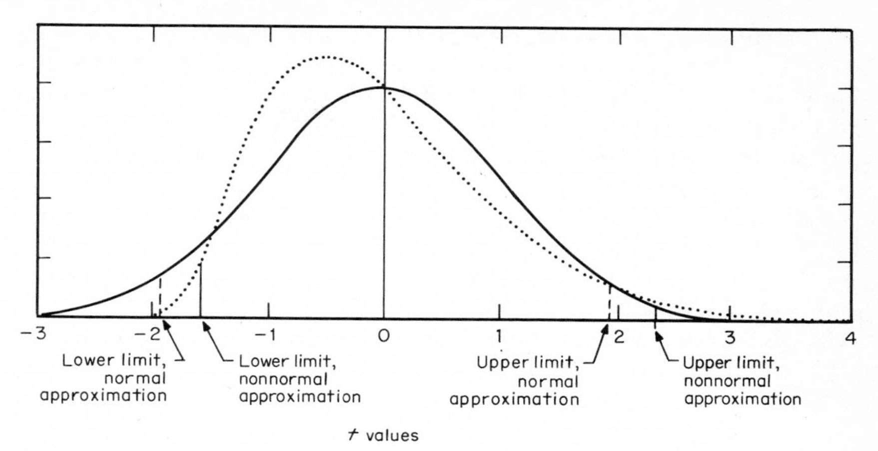

Fig. 3-1

clear that the usual normal approximation method provides a lower and upper limit both of which are too low.

The methods available to deal with this distortion are explained later.

The extent to which values tend to concentrate about the arithmetic mean as compared to the spread of the values in the tail of the distribution can be measured by a value $\alpha_4 - 3$. (See Technical Appendix 4 for the formula and method of computation.) This tendency of a distribution to exhibit either concentration about the average (arithmetic mean) or a more uniform spread is known as kurtosis. The value for $\alpha_4 - 3$ for a normal curve is zero. More strongly peaked or concentrated values as compared to the spread in the tails of the distribution are commonly found in all types of accounting data. For the distribution of invoices in Table 3-3, the value of $\alpha_4 - 3$ is 42, for the accounts receivable (Table 3-4) the value of $\alpha_4 - 3$ is 38 while for inventory values (Table 3-5) it is 13. The value of $\alpha_4 - 3$ for the errors found in the previously mentioned government audit was 15. This too will have an impact on the sampling distribution. However, the measure of kurtosis for a sampling distribution of the arithmetic means of samples drawn from a nonnormal distribution is:

$$(\alpha_4 - 3)_{\bar{x}} = \frac{\alpha_4 - 3}{n}$$

where $(\alpha_4 - 3)_{\bar{x}}$ = measure of kurtosis of sampling distribution
$\alpha_4 - 3$ = measure of kurtosis of population values
n = sample size

For samples of 100 or more, it can be seen that the measure of peakedness will be close to zero even when the value of $\alpha_4 - 3$ of the population is very large. Hence, the impact of the high concentration of values found in accounting data on the sampling distribution of the average and on the confidence limits will be limited, even in fairly extreme cases.

Since for many accounting populations, the confidence limits will be distorted by the above factors, it is necessary that appropriate steps be taken. Generally, the distortion will not be great in most audit sampling situations, even when dealing with a badly skewed and peaked population, providing a sample of *at least* 100 and preferably more units is used.

If the α_3 and α_4 values of the population are known or are estimated from the sample,[15] the true limits can be established by substituting, for the t values derived from the normal distribution, multipliers obtained from tables based on nonnormal distributions. Such values are provided in Table 3-6.

For instance, a sample of 100 is drawn from the invoice distribution after elimination of the highest 54 values (Table 3-3). The sample produces an arithmetic mean of $250.10 and a standard deviation of $670.00. The value of α_3 for the audited sample values turns out to be 5.9 and $\alpha_4 - 3$ equals 42.

The details of the method of calculation will be explored later, but at this point it can be determined that the confidence limits for the arithmetic average at the 90% confidence level will be a lower limit of $153.67 and an upper limit of $370.47. These values would compare to those obtained from the usual normal approximation of the sampling distribution which would produce a lower limit of $140.70 and an upper limit of $359.50.

In terms of the total value, based on the nonnormal sampling distribution, the confidence limit estimate is from $1,048,797.75 to $2,528,475.50 as compared with $960,277.50 to $2,453,587.50 for the normal approximation.

The point estimate of the population total is $1,706,933.00. It will be noted that the difference between the point estimate and the lower limit (a difference of $658,134.75) is *less* than the difference between the point estimate and the upper limit (a difference of $821,543.00) for the nonnormal confidence limits, by an appreciable amount.

[15]Since usually the population consists of the audited values, in most audit situations the values of α_3 and α_4 for audited values would have to be estimated from the sample.

Table 3-6a t ***(Standardized Deviates) Values for Mean of Samples from Nonnormal Populations***

[90% Confidence Level (Two-Sided)]
[95% Confidence Level (One-Sided)]

$\frac{\alpha_3^2}{n}$ (skewness)	$\frac{\alpha_4 - 3}{n}$ (kurtosis)				
	−0.2	0.0	+0.2	+0.4	+0.6
			Lower-Limit Deviates		
0.00	−1.65	−1.645	−1.64	−1.64	−1.63
0.01	−1.62	−1.62	−1.61	−1.61	−1.61
0.03	−1.59	−1.59	−1.59	−1.59	−1.59
0.05	−1.57	−1.58	−1.58	−1.58	−1.58
0.10	−1.54	−1.55	−1.55	−1.55	−1.55
0.15	−1.51	−1.52	−1.53	−1.53	−1.53
0.20	−1.48	−1.49	−1.50	−1.51	−1.52
0.30	−1.42	−1.44	−1.46	−1.47	−1.48
0.40	−1.35	−1.39	−1.42	−1.43	−1.45
			Upper-Limit Deviates		
0.00	+1.65	+1.645	+1.64	+1.64	+1.63
0.01	+1.68	+1.67	+1.67	+1.66	+1.65
0.03	+1.70	+1.69	+1.69	+1.68	+1.67
0.05	+1.72	+1.71	+1.70	+1.69	+1.68
0.10	+1.75	+1.74	+1.73	+1.72	+1.71
0.15	+1.77	+1.76	+1.75	+1.74	+1.73
0.20	+1.80	+1.78	+1.77	+1.76	+1.74
0.30	+1.84	+1.82	+1.80	+1.79	+1.77
0.40	+1.88	+1.86	+1.84	+1.82	+1.80

SOURCE: Values extracted from E. S. Pearson and H. O. Hartley, *Biometrika Tables for Statisticians*, vol. I, Cambridge University Press, New York, 1954, table 42.

NOTE: The values of α_3 and α_4 are for the sampled population although usually estimated from the sample.

Table 3-6b ***t (Standardized Deviates) Values for Means of Samples from Nonnormal Populations***

[*95% Confidence Level (Two-Sided)*]
[*97.5% Confidence Level (One-Sided)*]

$\frac{\alpha_3^2}{n}$ (skewness)	$\frac{\alpha_4 - 3}{n}$ (kurtosis)				
	−0.2	0.0	+0.2	+0.4	+0.6
	Lower-Limit Deviates				
0.00	−1.94	−1.96	−1.97	−1.98	−1.99
0.01	−1.89	−1.91	−1.93	−1.94	−1.95
0.03	−1.85	−1.87	−1.89	−1.90	−1.91
0.05	−1.82	−1.84	−1.86	−1.88	−1.89
0.10	−1.76	−1.79	−1.81	−1.83	−1.85
0.15	−1.70	−1.74	−1.77	−1.79	−1.81
0.20	−1.65	−1.69	−1.72	−1.75	−1.77
0.30	−1.55	−1.60	−1.65	−1.68	−1.71
0.40	−1.45	−1.52	−1.57	−1.61	−1.65
	Upper-Limit Deviates				
0.00	+1.94	+1.96	+1.97	+1.98	+1.99
0.01	+1.99	+2.01	+2.02	+2.02	+2.01
0.03	+2.03	+2.04	+2.05	+2.05	+2.05
0.05	+2.05	+2.06	+2.07	+2.11	+2.07
0.10	+2.09	+2.10	+2.11	+2.14	+2.11
0.15	+2.13	+2.13	+2.14	+2.16	+2.14
0.20	+2.15	+2.16	+2.16	+2.21	+2.16
0.30	+2.20	+2.21	+2.21	+2.25	+2.20
0.40	+2.24	+2.25	+2.25	+2.28	+2.24

Source: Values extracted from E. S. Pearson and H. O. Hartley, *Biometrika Tables for Statisticians*, vol. I, Cambridge University Press, New York, 1954, table 42.

NOTE: The values of α_3 and α_4 are for the sampled population although usually estimated from the sample.

Table 3-6c **t *(Standardized Deviates) Values for Means of Samples from Nonnormal Populations***

[*99% Confidence Level (Two-Sided)*]
[*99.5% Confidence Level (One-Sided)*]

$\frac{\alpha_3^2}{n}$ (skewness)	$\frac{\alpha_4 - 3}{n}$ (kurtosis)				
	−0.2	0.0	+0.2	+0.4	+0.6
	Lower-Limit Deviates				
0.00	−2.49	−2.58	−2.65	−2.71	−2.76
0.01	−2.38	−2.48	−2.55	−2.61	−2.67
0.03	−2.30	−2.39	−2.48	−2.54	−2.60
0.05	−2.23	−2.33	−2.42	−2.48	−2.54
0.10	−2.10	−2.21	−2.30	−2.38	−2.44
0.15	−1.99	−2.11	−2.20	−2.28	−2.35
0.20	−1.89	−2.01	−2.11	−2.20	−2.27
0.30	−1.71	−1.84	−1.95	−2.04	−2.13
0.40	−1.55	−1.68	−1.79	−1.90	−1.99
	Upper-Limit Deviates				
0.00	+2.49	+2.58	+2.65	+2.71	+2.76
0.01	+2.58	+2.66	+2.73	+2.79	+2.84
0.03	+2.64	+2.72	+2.79	+2.85	+2.93
0.05	+2.68	+2.76	+2.83	+2.88	+2.99
0.10	+2.73	+2.82	+2.89	+2.95	+3.03
0.15	+2.77	+2.86	+2.93	+2.99	+3.07
0.20	+2.80	+2.89	+2.96	+3.02	+3.12
0.30	+2.83	+2.93	+3.01	+3.07	+3.16
0.40	+2.84	+2.95	+3.04	+3.11	+3.19

Source: Values extracted from E. S. Pearson and H. O. Hartley, *Biometrika Tables for Statisticians*, vol. I, Cambridge University Press, New York, 1954, table 52.

NOTE: The value of α_3 and α_4 are for the sampled population although usually estimated from the sample.

METHODS FOR CONFIDENCE LIMITS FOR SAMPLES FROM SKEWED POPULATIONS

When a sample is drawn from a badly skewed population, it is evident that the confidence limits computed using the customary normal approximation will not be precisely correct.

There are several methods available for dealing with this situation. The first method is the simplest but frequently not the most practical. The method merely requires that the sample be large enough to assure a sampling distribution which will be reasonably normal.[16]

Unfortunately, for very badly skewed distributions ($\alpha_3 = 5$) the resulting sample size may be about 400 or more. For moderately skewed distributions, such as the inventory value distribution, 125 or more would serve. Since the use of samples of 400 or more in the audit situation is not common, other approaches would be required.

Another possible approach would be to recognize the distortion, no matter what sample size is used, and to allow for it. This method is particularly useful when a claim is to be made for payment or repayment, as when disallowances are established. If, as suggested earlier (pages 58 to 59), the lower limit computed by this normal approximation method is used to determine the demand for payment, this certainly is a fair basis for stating that at least this amount is due. As seen above, this normal-approximation lower limit understates the true lower limit when sampling from positively skewed distributions. If adjustments or errors which result in payment are considered as positive values and the point estimate is positive, the population of disallowances and the resulting sampling distribution of averages will be positively skewed as well.

The use of such a lower limit results in an advantage for the individual or organization being audited. However, cost and time limitations may make such a lower figure satisfactory. It permits the use of relatively small samples which nevertheless produce defensible though conservative results.

The next and most complex method is to compute the precise confidence limits using values of "t" in the formula

$$SE_{\bar{x}} = t\frac{\sigma}{\sqrt{n}}\sqrt{1-\frac{n}{N}}$$

obtained from a special table providing such values for nonnormal distributions. A set of such "t" values are included in Table 3-6 for a range

[16]The criterion that was used here was that the sampling error be off by not more than 5%.

of values likely to be encountered in audit situations. An explanation as to the manner in which these values were obtained can be found in Technical Appendix 5.

The difficulty with this method is that it requires the computation of the measure of skewness (α_3) and kurtosis ($\sigma_4 - 3$) from either the population or sample, a difficult computation. However, if performed on a computer or with an electronic calculator it may not be impractical. The problems and methods of such computations are discussed in Technical Appendix 4.

The value of "t" to be used to replace that obtained from the normal distribution is obtained using the α_3 and α_4 of the population estimated from the sample with the entry to the table being

$$\frac{\alpha_4 - 3}{n} \quad \text{for kurtosis}$$

and

$$\frac{\alpha_3^2}{n} \quad \text{for skewness}$$

These values represent the $\alpha_4 - 3$ and the α_3^2 of the resulting sampling distribution. (See Technical Appendix 5 for an explanation.)

It should be noted that each confidence limit must be separately computed, since there is a different value of "t" for the lower limit as compared to the upper limit. To obtain the lower limit, the precision obtained using the adjusted t value which is obtained by reference to the value of α_3 and α_4 estimated from the sample is subtracted from the point estimate. For the upper limit, the new precision is added to the point estimate.

It will be noted that the value of "t" for lower limit computation is always about the same as or less than the normal curve value in the range of α_3 and α_4 values involved. This means that the precise lower confidence limit will be higher than that obtained by the normal approximation as previously noted. For instance, in the 90% confidence level (two-sided) table all of the values are equal to or greater than 1.645.

On the other hand, the upper limit deviates ("t") are about the same or higher than the 1.645 of the normal approximation.

Referring to the example on page 66, the computation of the confidence limits can now be explained. The facts given about the invoices in Table 3-3 were as follows after elimination of the 54 highest values for 100% audit:

$$N = 6825$$
$$n = 100$$

$$\overline{X} = 250.10$$
$$\sigma = 670.00$$
$$\alpha_3 = 5.9$$
$$\alpha_4 - 3 = 40$$

Reference to Table 3-6 provides a "t" value (by interpolation) for the lower limit of -1.45 and $+1.81$ for the upper limit.

The point estimate of the total is

$$N\overline{X} = 6825(\$250.10) = \$1{,}706{,}932.50$$

The sampling error of the average for the lower limit is

$$SE_{\bar{x}} = t\frac{\sigma}{\sqrt{n}}\sqrt{1 - \frac{n}{N}}$$

$$= -1.45\,\frac{\$670}{\sqrt{100}}\sqrt{1 - \frac{100}{6825}}$$

$$= -1.45(\$67)(.9926)$$

$$= -\$96.43$$

and for the total

$$NSE_{\bar{x}} = -\$96.43(6825) = -\$658{,}142.19$$

The lower limit for the total is

$$\$1{,}706{,}932.50 - \$658{,}142.19 = \$1{,}048{,}797.75$$

The sampling error of the average for the upper limit is

$$+1.81\,\frac{\$670}{100}\sqrt{1 - \frac{100}{6825}} = +\$120.37$$

and for the total

$$NSE_{\bar{x}} = +\$120.37(6825) = \$821{,}543.01$$

and for the upper confidence limit

$$\$1{,}706{,}932.50 + \$821{,}543.01 = \$2{,}528{,}475.51$$

A note of caution must be observed here. In the above example, the distribution was of *unaudited* values, and α_3 and α_4 could be computed from the population values.

In most audit situations, the projection to be made is of the *audited* values or errors. The population of these values is unknown and it will be necessary to calculate α_3 and α_4 from the sample *audited* values as an estimate of that of the population in the same manner as is customary for the standard deviation.

Another approach to overcoming the difficulties arising when sampling values in nonnormal distributions commonly encountered for accounting data has been suggested by Burns and Loebbecke.[17] This is a very conservative technique which uses the value determined by the Tchebycheff inequality for the value of t in the computation of the sampling error.

The Tchebycheff inequality indicates that regardless of the nature of the distribution of the data involved, the percentage of the observations included in a range of k standard deviations measured above and below the arithmetic mean will be at least

$$1 - \frac{1}{k^2}$$

By use of this formula, it is possible to determine the *minimum* proportion of the averages (or totals) of samples in a sampling distribution which will be included within a given sampling error measured on both sides of the arithmetic mean, regardless of the actual type of sampling distribution involved, when the sampling error is computed using $t = k$.

The value of k required to achieve *at least* a probability (confidence) level of including the population value (arithmetic mean or total) within confidence limits for *any* population can be computed from the above formula as shown below.[18]

Confidence Level, %	k
90	3.2
95	4.5
99	10.0

It is to be emphasized that the probabilities specified by this method as the confidence levels are the *minimum* probabilities and not the pre-

[17] D. C. Burns and J. K. Loebbecke, "Internal Control Evaluation: How the Computer Can Help," *Journal of Accountancy*, August 1975, p. 68.

[18] The value of k can also be computed for any confidence level from:

$$k = \sqrt{\frac{1}{1 - \text{CL}}} \quad \text{where CL} = \text{confidence level}$$

cise confidence levels. The exact confidence levels will be determined by the distribution of the values sampled, but in almost all cases the actual confidence levels will be appreciably *higher* than those specified by this method.

Thus, this is an *extremely* conservative method which will require a much larger sample than that required for the previously discussed methods in order to achieve a stated confidence level or will result in a much larger sampling error for a given sample size.

As can be seen by reference to Table 3-6, these values of k, to be used to replace t in the computation of the sampling error, are much larger than the t values to be used for any of the distributions likely to be encountered in accounting data when the data are truncated (stratified) appropriately.

However, when it is not desired to calculate the value of the measures of skewness (α_3) or kurtosis (α_4) required by the previously discussed method, or where it is desired to be absolutely certain of achieving *at least* the stated confidence level, the method would be appropriate, although expensive.

Chapter 4

Inventories and Sampling

The valuation of the inventory of a business results is a key figure in the financial statements of the company. If material, the inventory value may be critical in determining both the profit and net worth of a company.

Yet in many situations it is difficult, if not impossible, to establish a precise value for an inventory on hand at a given date. A considerable part of the problem may arise from the inventory taking process itself, as there may be errors in identification, counting, pricing, and possibly in computing and footing the extensions.

When perpetual inventory records are maintained, the actual counts are the result of a complex of receipts, withdrawals, shrinkage, and pilferage and counts often do not agree with the recorded figures. Such records are seldom even close to absolute accuracy.

While it is customary for management to verify perpetual inventory records or to establish balances on hand by 100% physical inventories, experience has shown that such counts cannot be completely trusted because of the virtually invariable inaccuracies that arise in large inventories.

Traditionally, the recorded inventory value has acquired an illusion of absolute accuracy, since it arises from a purported 100% count. However, it must be remembered that the inventory determination of quantities often involves approximations. Counts of numerous items in bins are often approximated by weighing or by other estimation methods. Inventories of large quantities of packaged goods will not usually require that all, or even many, be opened to ascertain that every unit is actually contained within each multi-unit package. Many thousands of loose items, such as cigarettes or small parts on a factory floor, will not be individually counted or even entirely weighed to obtain a count. Supposedly reasonable approximations may be used.

Of even greater importance is the fact that since physical inventories ordinarily are taken but once a year, it is not practical to maintain year-round staffs of inventory takers. Generally a large part of the available staff must be impressed into service for a day or two to achieve such inventorying. Because of the involvement of some, if not many, who may not be thoroughly familiar with every item in the inventory, combined with a general lack of enthusiasm for a task not directly related to their usual activities and of a tiring and boring nature, errors of count and identification frequently arise.

In many situations, the problem may well be resolved by merely keeping the errors within reasonable bounds.

It is not intended to imply that all, or even many, inventory values contained in financial statements are grossly inaccurate. Nevertheless, the point must be made that a physical inventory usually gives rise to an *estimate* of the inventory value and is not a precise figure, accurate to the nearest penny.

Statistical sampling may play an important role with respect to this problem and in other areas related to inventories. The sample provides an ability to make an estimate of the total inventory value with a limited effort. The difference between the *estimate* obtained from a 100% physical inventory and the *estimate* projected from a statistical sample is of considerable consequence. The 100% physical inventory estimate includes the impact of unknown errors in counting, pricing, extension, and summarization. On the other hand, the much smaller task involved in obtaining the sample projection provides an ability to achieve a high degree of accuracy for the limited number of items included in the sample. The sample projection usually involves only a *known* sampling error. The limited sample size enables the use of only the best qualified personnel, thus eliminating errors of item identification.

Some of the applications of statistical sampling in the inventory area include:

1. Observation of inventories
2. Valuation of inventories without 100% physical inventories
3. Reevaluation of inventories on a different price basis

INVENTORY OBSERVATION

The AICPA has established inventory observation as a generally accepted auditing procedure. *SAS 1* states:

> Confirmation of receivables and observation of inventories are generally accepted auditing procedures. The auditor who issues an opinion when he has not employed them must bear in mind that he has the burden of

> justifying the opinion expressed. [It is further noted that] it is ordinarily necessary for the independent auditor to be present at the time of the count and, by suitable observation, tests and inquiries, satisfy himself respective to the effectiveness of the method of inventory taking and the measure of reliance which may be placed upon the client's representation about the quantities and physical condition of the inventories.[1]

It is emphasized that "when the independent auditor has not satisfied himself as to the inventories in the possession of the client through the procedures described . . . , tests of the accounting records alone will not be sufficient for him to be satisfied as to quantities; it will always be necessary for the auditor to make or observe, some physical counts of the inventory . . ."

In a description of the auditor's responsibility in this area W. G. Patten states:

> The independent auditor has a primary responsibility to ascertain, insofar as practical, that the amount shown for inventories is represented by physical goods and that reasonable care has been taken in determining the physical quantities and their condition. He must ascertain that the quantities have been fairly and consistently priced in accordance with generally accepted accounting principles and that they are carefully extended and summarized. He must satisfy himself that they include no obsolete or defective goods except at fair values and that due provision has been made for probable losses on slow moving goods. . . . [2]

A large part of the compliance with this responsibility is often accomplished by investigation of the physical inventory *process*. Customarily this observation includes an examination of the inventory taking system, especially with respect to internal controls. Frequently this observation includes some tests of counting and pricing and, possibly, of summarization.

However, prior to the advent of the statistical sampling approach, there was no way of directly translating the results of such tests of counts and pricing into a determination that the reported inventory value was reasonable. Of course, detection of widespread errors in counting and pricing might result in a demand that the inventory be retaken, but without statistical sampling there would be no way of using such results to ascertain the actual inventory value.

The application of estimation sampling for variables (dollar value sampling) provides a solution to the problem of determining the impact of the errors in inventory taking upon the reported inventory value.

[1] *Statement on Auditing Standards 1,* p. 58.

[2] W. G. Patten, "Inventories," in J. A. Cashin (ed.), *Handbook for Auditors,* McGraw-Hill Book Company, New York, 1971, pp. 22–24.

When using the statistical sampling technique, the auditor performs a count for a probability sample of inventory items, prices these items independently, extends the results, and projects an interval estimate of the total dollar value of the inventory.

If the reported inventory value falls within the projected interval estimate it *cannot* be concluded that there is a material error in the stated book record value, provided that the total spread of the interval estimate does not exceed the minimum amount considered to be material.

The interval estimate of the total audited (true) value of an inventory is based on the point estimate derived from the test sample with an allowance for the sampling error. The interval estimate obtained from the sample has a probability equal to the confidence level of including within its limits the audited value of the inventory that would be obtained by a 100% audit.

If the book record value is contained within the limits of the interval estimate, it can be no further from the true audited value than the width of the interval estimate. Thus, if the distance between the upper and lower limits of the confidence interval does not exceed the minimum amount considered to be material, the probability that there is a material error is very small.

Therefore, as previously suggested, if the sample precision achieved is equal to one-half of the minimum amount considered to be material the total spread of the interval will not be more than that amount. Under this condition, if the book record value is included within the confidence interval estimate, most probably the inventory value is not materially misstated. This conclusion may be reached since with the book record value contained between the two confidence limits, the difference between the farthest limit and the book record value cannot exceed the minimum amount considered to be material (*MM*) because the entire width of the interval cannot exceed that amount.

If the book record value is contained within the interval but the spread of the interval is greater than the minimum amount considered to be material, there is no evidence that a material error exists, though it may.

If the book record value is *not* included within the confidence interval estimate, there is a high probability that a *real* difference (error) exists between the book record value of the inventory and its true or audited value. However, this difference *may or may not* be material. It may be concluded that the true difference (error) is *at least* equal to the difference between the book record value and the *nearest* limit of the interval estimate.

Since there is a high probability that the book record inventory value will be included within the interval estimate when correct, a book record value outside of this interval indicates a high probability of a disparity between the two values. If a 95% confidence level has been used there is only a 5% probability that the book record value is actually outside of the interval, although correct.

If the book record value is outside of the confidence interval estimate of the total inventory value projected from the sample *and* the difference between the point estimate of the total value and the book record value exceeds the minimum error amount considered to be material (MM), there is a high probability that a material error exists, especially if the sampling error is $\frac{1}{2}MM$ or less.

The reasoning leading to this conclusion may be seen by reference to Figure 4-1 where the inventory book record value is assumed to be overstated by exactly the minimum amount considered to be material.

The concept of the sampling distribution asserts that the point estimates of the total inventory values from all possible samples are clustered about the "true" (audited) inventory value as shown in that figure.

There is a probability equivalent to the confidence level that the projection (point estimate) obtained from the audit sample actually will be within the confidence limits (upper and lower) about the true value, equal to the confidence level used to establish that interval.

If, as in Figure 4-1, a sample precision equal to one-half of the minimum amount considered to be material ($\frac{1}{2}MM$) is achieved and a material error of overstatement actually exists, it is apparent that the audit sample will produce a point estimate which will most likely not exceed the upper limit of the interval estimate and that in turn the upper limit of the confidence interval produced by such a sample will not exceed the book record value. Hence, the book record value of such a materially overstated inventory will not be included in the sample confidence interval estimate, meeting the first requirement to establish the existence of a material error. However, the second requirement is

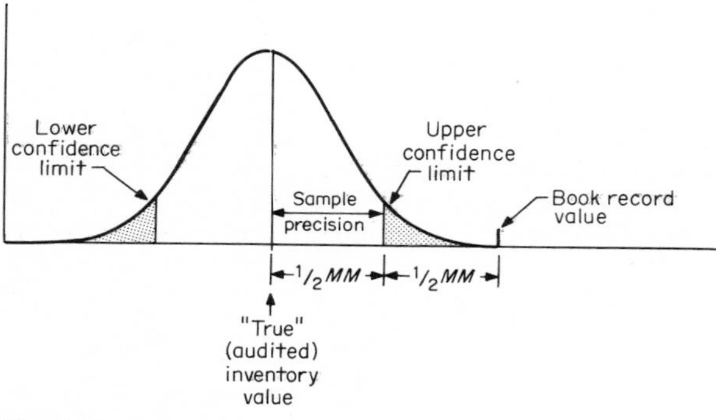

Fig. 4-1

that the sample point estimate differ from the book record value by at least the minimum error amount considered to be material (MM). It can be seen from Figure 4-1 that if an error of overstatement greater than the minimum amount considered to be material actually exists, it is likely that this requirement will be met as well.

However, the probability that such a disparity will be declared to be material (meet both requirements) is a function of the actual difference *and* the confidence level selected as well as the sample precision actually achieved. The probability of detecting a material error of overstatement (or understatement), when such an error is equal to the minimum amount considered to be material or is some multiple of that value for various confidence levels, sample precisions, and amounts of error (as a multiple of the minimum amount considered to be material), is shown in Table 4-1.

The methods of evaluating sample projections of inventory values described above are summarized below.

1. If the book record value is included between the upper and lower confidence limits *and:*

  a. The sampling error actually achieved is not more than one-half of the minimum amount considered to be material ($\frac{1}{2}MM$), there is a high probability (equal to the confidence level) that no material error exists.

  b. The sampling error actually achieved is greater than one-half of the minimum amount considered to be material, there is no evidence that there is an error in the book record value and thus there is no conclusive evidence that a material error exists, although it may.

2. If the book record value of the inventory is *not* included within the confidence interval estimate *and:*

  a. The difference between the point estimate of the total inventory value and the book record value *is equal to or greater* than the minimum error amount considered to be material, it may be concluded that a material error probably exists.

  b. The difference between the point estimate and the book record value is *less* than the minimum error amount considered to be material (MM), there is a high probability that the total book record inventory value is in error but the error may or may not be material.

The method of arriving at these probabilities can be explained by reference to Figure 4-1.

Table 4-1 ***Probability of Error Detection and Declaration of Material Error When Actual Error Is Equal to* MM *or Some Multiples of* MM *for Various Confidence Levels and Sample Precisions***

Confidence Level, %, and Sample Precision	Probability: Of Error Detection, %*	Probability: Declaration of Material Error, %†
Exactly Minimum Material Error Amount		
Confidence level = 90 Sample precision = $M/2$	95	50
Confidence level = 95 Sample precision = $M/2$	97.5	50
Confidence level = 90 Sample precision = $M/3$	99.9	95
Confidence level = 95 Sample precision = $M/3$	99.9	97.5
$1\frac{1}{2}$ Times Material Amount ($1\frac{1}{2}MM$)		
Confidence level = 90 Sample precision = $M/2$	99.5	95
Confidence level = 95 Sample Precision = $M/2$	99.9	97.5
Confidence level = 90 Sample precision = $M/3$	99.9	99.9
Confidence level = 95 Sample precision = $M/3$	99.9	99.9
2 Times Material Amount ($2MM$)		
Confidence level = 90 Sample precision = $M/2$	99.9	99.9
Confidence level = 95 Sample precision = $M/2$	99.9	99.9
Confidence level = 90 Sample precision = $M/3$	99.9	99.9
Confidence level = 95 Sample precision = $M/3$	99.9	99.9

*Probability that book record value will be outside the confidence interval (one-sided).
†Probability that the difference between the point estimate and the book record value would exceed the minimum amount considered to be material *and* the book record value is outside the confidence interval.

For instance, for the first entry in Table 4-1, where the confidence level specified is 90% and a sampling error of the total equals $\frac{1}{2}MM$, with an actual error exactly equal to MM, the probability that an error would be indicated is at least 95%. This arises from the fact that for an overstatement with a sample point estimate of at most equal to the upper limit of the sampling distribution about the true value, the confidence interval derived from that sample would provide an upper limit exactly equal to the book record value. The probability that a sample with a higher point estimate, thus producing a higher upper limit in such circumstances, is

$$\frac{1 - \text{CL}}{2}$$

where CL = confidence level.

Hence, the probability of indicating an overstatement error of any magnitude would be equal to 1–05. The same reasoning applies to errors of understatement.

To meet the requirement to declare the error material in amount it is necessary that the point estimate differ from the book record value by at least the minimum amount considered to be material (MM). Reference to Figure 4-1 indicates that such a sample point estimate would have to be less than the "true" (audited) value. It can be seen that only 50% of the samples would produce such low point estimates for the given specifications. Thus, the probability that a disparity would be declared to be material is 50%, as indicated in Table 4-1.

However, there is another risk involved. That risk relates to the probability that an error will be indicated when there is none or that a material error will be declared when there is none.

If, as suggested previously, the sample precision actually achieved is equal to one-half of the minimum amount considered to be material (MM), the probability that an error will be indicated where there is none will be the complement of the confidence level. If a 95% confidence level is used, the probability of falsely indicating an error where none actually exists is 5% (100% − 95%).

However, when such a false indication of an error occurs, it does not of necessity result in the conclusion that there is a material error. For an error to be declared material, the point estimate must also differ from the book record value by the minimum amount considered to be material. If a sample precision equal to one-half of the minimum amount to be considered to be material has been achieved and a reasonable confidence level has been used, the probability of a false indication of a material error when the book record value is correct is virtually nil.

Table 4-2 ***Probability of False Indication of Material Overstatement***

Confidence Level	Sample Precision Achieved	Probability, %, When Actual Error Is	
		None	$\frac{1}{2}MM$
90%	$\frac{1}{2}MM$	0.05	5.0
	MM	5.0	20.0
95%	$\frac{1}{2}MM$	0.005	2.5
	MM	2.5	16.0

The probability that a false indication of a material error will take place is a function of the actual disparity, the confidence level used, and the sample precision achieved.

To declare that a material error exists in the inventory value the point estimate of the sample must differ from the book record value by MM and the book record value must be outside the interval estimate. If the book record value is exactly correct and a sample precision of $\frac{1}{2}MM$ is achieved, the probability that the point estimate will differ from the book value by more than MM and indicate a material error is less than 0.01%.

However, if the sample precision achieved in such a case is poorer, say equal to the minimum amount considered to be material, the probability becomes 2.5% for a 95% confidence level.

In addition, when the true (audited) value actually differs from the book record value by some amount less than that considered to be material, the probability of a false indication of a material error increases.

Table 4-2 indicates the risk of a false indication of a material error for various actual errors, confidence levels, and sample precisions.

An examination of Table 4-2 indicates that it is necessary to have a sample precision of close to $\frac{1}{2}MM$ in order to avoid an appreciable probability of a false indication of a material error and the resulting unnecessary increase in audit effort and cost.

INVENTORY SAMPLING PROBLEMS

There are a number of unique problems which arise when sampling to confirm the reasonableness of an inventory value.

In order to select the necessary sample, the sampling unit must be established and the population defined. Ideally, the sampling unit would be each stock item identification, such as a stock item number. To accomplish a selection on this basis, however, a complete and accurate listing of all stock items must be available. Such a listing would comprise the *frame*, in statistical terminology. If not all stock items are on the listing, they cannot be included in the sample. The frame and the population will not be the same. As a result the projected value will be that for the items in the frame only. If the listing contains items of which there are not quantities on hand, no harm is done since a value of zero will be recorded for any such items in the sample. If the same items are listed more than once but it is not apparent that there are such multiple listings, the projection will result in an overestimate.

If, as is frequently the case, the sampling unit is defined to be inventory tags or cards, these problems can become serious. Merely sampling inventory tags will provide unreliable projections without some assurance that all tags are available to be sampled, that there are no duplicates, and that the counts are accurate—representing merchandise actually on hand.

Merely sampling inventory tags without confirming the counts on them is an invitation to the fraud-minded to create tags for nonexistent stock or to exaggerate the counts. Mere observation of the inventory process and evaluation of the system do not prevent fictitious counts.

The level at which the auditor starts his test will determine the extent and validity of the test. Thus, if tags are sampled without a confirmation of the count, it must be recognized that the client's count is automatically accepted.

If the sampling is made from a listing of the price-extended values, it must be recognized that *both* the count *and* the pricing by the client have been accepted and that the test will merely establish the reasonableness of the summarization.

The ideal approach is for the auditor to sample from a listing of stock items known to be complete, independently count and verify the pricing for the sample items, and make a projection from the results in order to test the reasonableness of the reported book record value.

Other sampling units may in some cases be more appropriate. In one instance, the sampling unit was a "lot" where the lot was defined as "any group of identical parts or assemblies which are physically touching." These lots were defined by placing a slip containing a number on each lot. The extent of the lot was indicated by a cord placed around the group of items defined as a lot.[3]

[3]A. L. Rudell, "Applied Statistical Sampling Doubles Inventory Accuracy, Halves Costs." *NAA Bulletin*, October 1957, sec. I, pp. 3–11.

Another approach to the definition of a sampling unit consisted of theoretically dividing a floor of a large factory into small areas with assigned numbers. A random selection of these areas was inventoried and the results projected. In this case, the sampling unit was an area and the population consisted of all areas on the floor.

These last two methods have the advantage of giving assurance that everything in the inventory is covered and given an opportunity of being included in the sample. Dependence on a defective stock item listing or on inventory tags does not provide such assurance.

In general, the definition of the sampling unit for inventory sampling will be a function of the situation encountered such as the records available and the physical distribution of the items to be inventoried.

However, since the sample size required to achieve a given precision is *not* closely related to the population size, the greatest savings will result when the sampling unit is small and the population large, if this does not greatly increase the variability of the values involved. Further, the cost of the actual counting will be reduced when the quantity is limited by using smaller sampling units.

VALUATION OF INVENTORIES

In addition to inventory observation, there are other useful applications of statistical sampling in determining the total value of an inventory without 100% physical inventorying. Such sampling plans can be designed and have been used to provide an inventory value for use in financial statements in lieu of the usual 100% physical inventory. These methods are particularly useful in providing inventory values at other than year-end periods, when such values are needed.

In recognition of the importance of the inventory value in the financial statements, it is customary to take a physical inventory each year to fix the value of the inventory on hand by actual count and valuation as well as to adjust perpetual inventory records.

The importance of and necessity for periodic inventories cannot be denied. However, the accomplishment of inventory counts is costly and disturbing, often involving large numbers of persons working overtime or on weekends. When the cost of such an inventory is computed, in some instances it is a significant percentage of the inventory value and in other cases a large and important expense. Further, as previously noted, the results of such 100% counts are often far from accurate.

The use of a probability sample as a substitute for a 100% inventory, where a very limited number of sampling units are subject to count, can result in a much lower cost and in increased accuracy. Assigning the best informed personnel to the count of the sample items together with the possibility of a careful check or recount of the limited number of items involved can assure accuracy.

SAS 1 states: "In recent years, some companies have developed inventory controls or methods of determining inventories, including statistical sampling, which are highly effective in determining inventory quantities and which are sufficiently reliable to make unnecessary the physical count of each item of inventory."[4]

When the inventory value is projected from the sample for use in the financial statements, the value to be used is the point estimate.[5] However, it should be apparent that to obtain a useful value for this purpose, the sample must be large enough to provide an adequate sample precision at a reasonably high confidence level.

The required precision for this purpose should be much less than the minimum amount considered to be material divided by two ($\frac{1}{2}MM$) as previously suggested. While such a precision will protect against the possibility of introducing a material error, it generally cannot be considered adequate.

The sample precision required will be a function of the situation encountered as well as the materiality of the inventory value itself. If the total inventory amount is not considered to be material in the financial statements, a very close precision may not be required. However, when the total value of the inventory is critical to the financial statement, a sampling precision of one-fifth or one-tenth of the minimum amount considered to be material may be appropriate. The confidence level for such purposes would be 95% or higher.

REEVALUATION OF INVENTORIES

Statistical sampling can play a particularly useful role in the reevaluation of inventories on something other than the original pricing basis. An example would be the conversion from first-in–first-out (FIFO) to last-in–first-out (LIFO) or establishing the replacement cost of the inventory. The use of statistical sampling in conversions to LIFO in

[4]It is interesting to note that *SAS 1* also states: "If statistical sampling methods are used by the client in the taking of a physical inventory, the auditor must be satisfied that the statistical sampling plan has statistical validity, that it has been properly applied, and that the resulting precision and reliability as defined statistically, are reasonable in the circumstances." Both statements are from *Statement on Auditing Standards 1*, p. 61.

[5]The point estimate is obtained by multiplying the average dollar value of the sampling units by the population size for an unrestricted random sample.

relation to business taxes as covered in the Internal Revenue Service procedures is discussed in Chapter 6.

In such reevaluations of a past inventory, a sample of the items in the inventory are repriced on the new basis and the results projected. The considerations in this type of projection are the same as those described above for original evaluations. It must be remembered that such reevaluations reflect only the effect of the new pricing basis and in no way confirm the reasonableness of the count data in the original inventory.

Chapter 5

Confirmations

The publications of the AICPA indicate that a confirmation of accounts receivable should ordinarily be included in audits of commercial organizations. *SAS 1* states:

> Confirmation of receivables and observation of inventories are generally accepted auditing procedures. The independent auditor who issues an opinion when he has not employed them must bear in mind that he has the burden of justifying the opinion expressed.[1]

This audit philosophy has been recommended by the AICPA since 1939. In 1958 the Canadian Institute of Chartered Accountants issued a somewhat broader statement:

> [The] committee considers that some form of direct communication of accounts and notes receivable by communication with debtors is a useful auditing practice which has gained general acceptance.[2]

While the principal objective of the confirmation method is to establish the validity of the account balances, it also may be used to test compliance with internal controls and evaluate the collectibility of the accounts.

Confirmations are not restricted to receivables but are sometimes used in conjunction with payables, capital stocks, notes payable, etc. In each case the confirmation requires direct contact with individuals outside of the audited organization.

There are two forms of confirmation requests, positive and negative. Both of these methods entail communication with individuals or organizations involved in the account. In the case of confirmation of accounts or notes receivable the debtor is contacted. In the negative

[1] *Statement on Auditing Standards 1*, p. 58.

[2] *Confirmation of Receivables*, Bulletin No. 15, Canadian Institute of Chartered Accountants, Toronto, 1958, para. 4.

confirmation, the recipient is requested to reply only if he or she disagrees with the data supplied; in the positive method, a request is made for a response whether or not there is disagreement.

When positive confirmations are used, there is the implication that "a second and sometimes an additional request" should be made if no reply is secured.[3] Alternative procedures will be used when no response can be achieved.

There are unique sampling problems when the confirmations are selected on a sample basis, as they frequently are.

NEGATIVE CONFIRMATIONS

When negative confirmations are used, the only replies to be expected will be from recipients of the confirmation request who recognize errors *and* are willing to respond.

If a sample of 200 is drawn and 10 responses are received which indicate significant errors, it cannot be concluded that only 5% of the records for those 200 individuals in the sample included errors. Still less appropriate is such a conclusion for the population.

There will be some failures to respond, where errors do exist in the sample records, because of inability to recognize the errors or unwillingness to respond. If an error rate is computed based on the number responding and indicating significant errors, there is an automatic assumption that all of the other records were correct. This may not be true. Hence, there is no method of computing an error rate or the dollar value of errors among the sampled records with negative confirmations. Such a figure can be looked upon only as a *minimum* error amount, when projected to the population.

Since there is no way of determining the number who disregarded the confirmation request for one reason or another, there is no method of computing and projecting either the rate of error or the corresponding total dollar value of errors for the entire population.

The actual error rate indicated by the responses could be obtained only by dividing the reported number of significant errors by the total number who would have responded if there had been an error. This is an unknown value.

It may well be that some of the recipients of the confirmation request were *unable* to respond because of a lack of records or a simple inability to establish whether the account rendered is correct or incorrect. This may be particularly true in the case of retail customers.

[3] *Statement on Auditing Standards 1*, p. 60.

A situation of this sort may well cause a bias in the resulting data. Individuals who keep sound records or otherwise have an ability to recognize an error probably would have called attention to prior errors and received corrections, while the others would not. Negatives will then, to the extent such circumstances exist, understate the number and value of errors in the population.

Nevertheless, *SAS 1* recommends the use of negatives by stating that:

> The negative form is useful particularly where internal control surrounding accounts receivable is considered to be effective, when a large number of small balances are involved and when the auditor has no reason to believe that the persons receiving the requests are unlikely to give them consideration. If the negative rather than the positive form of confirmation is used, the number of requests sent or the extent of other auditing procedures used should normally be greater. . . .[4]

It must be emphasized that increasing the sample size does *not* overcome the deficiencies mentioned above and provides no better basis for projecting either the error rate or total dollar value of the errors.

If the purpose is merely to establish the nature of errors (if they exist in the population) or the existence of fictitious accounts by finding instances of such deviations in the sample, the technique to be used is discovery sampling. The objective here is to be reasonably sure of including at least one instance of such a deviation in the sample. Nevertheless, negative confirmations are still widely used, primarily on the grounds that they are less expensive, since no follow-up for non-respondents is required.

It is evident that projections from statistical samples relative either to the frequency or the dollar value of errors in the account balances will not be defensible when negative confirmations are used. Statistical sampling requires the use of positive confirmations.

A study group of the Canadian Institute of Chartered Accountants recommends that "auditors use the positive, and not the negative, method of communication with debtors to obtain the evidence necessary to support an expression of an opinion as to the authenticity of accounts receivable."[5]

POSITIVE CONFIRMATIONS

While the same problem of failing to receive replies exists for positive confirmations, the situation is different in some respects. Assume a

[4] *Statement on Auditing Standards 1*, pp. 89–90.

[5] *Confirmation of Accounts Receivable*, Study Group on Audit Techniques, Canadian Institute of Chartered Accountants, Toronto, 1971.

mailing of a sample of 200 confirmations where 100 replies were received and 10 significant errors reported. This indicates a 10 percent error rate *among those responding*. This is a calculable rate of error but is not necessarily projectable to the true error rate in the entire population.

Some auditors would respond to this situation by mailing an additional 200 confirmations, intending to obtain a total of 200 responses. Other auditors might anticipate such a result and oversample originally by mailing 400 requests to obtain the required 200 responses.

Neither of these actions would improve the results. Any computation of error rates or projection of dollar values from the resulting sample of responses would apply only to those in the population who would have responded to such a request. There is no reason to assume that the errors or deviations for those who would not respond are the same in frequency or amount as for those who would.

Loebbecke and Neter observe that: "We believe that it would be folly for the auditor to assume at this time that respondents to positive confirmations are like non-respondents."[6]

Little information is available on the response rate for positive confirmations, but undoubtedly such rates will vary widely with the type and characteristics of the business and the nature of the recipients. The rate of response to first requests may be quite low but usually will be appreciably increased by the usual second requests.

The AICPA implies a requirement that there be second requests when positives are used. *SAS 1* states:

> When the independent auditor sets out to confirm receivables by means of positive requests, he should generally follow up with a second and sometimes an additional request to those debtors from which he has received no reply.
>
> For important accounts involving large balances or which require attention for other reasons and for which no response has been secured in spite of a follow-up request, a telephone call requesting a response may well be successful.[7]

In an unpublished study of accounts receivable balances, using a probability sample, 4000 confirmation requests were sent to customers of a large public utility in a very large city. Responses to the first request constituted 34 percent of the mailing; an additional 30 percent responded to a second request, for a total reply rate of 64 percent. Other studies seem to indicate that overall response rates to both first and sec-

[6] J. K. Loebbecke and J. Neter, "Statistical Sampling in Confirming Receivables," *Journal of Accountancy*, June 1973, p. 46.

[7] *Statement on Auditing Standards 1*, para. 331.08.

ond requests for various business organizations may run from 60 to 90 percent.[8]

While it is difficult to generalize about response rates for all types of business organizations, it is clear that even after second requests there will be an appreciable rate of nonresponse.

Again, the reason for failing to respond to confirmation requests may be the inability or unwillingness of the recipient of the request to respond. In the accounts receivable situation, there are some organizations, such as governmental units, from which there will be no response either as a matter of policy or because the nature of the accounting system renders such responses difficult or impossible.

Several techniques are available to increase the response rate. One method for use with accounts receivable is to send requests relative to open items or invoices rather than account balances. Thus the sampling unit becomes the invoice (or open item) rather than the account balance. Individuals or organizations may find it easier to respond to such a request. Some organizations which routinely refuse to respond to account balance confirmation requests will often respond to an inquiry about an invoice.

The change in the sampling unit will have no impact on the probability sampling procedures other than to change the population size. The projection obtained from a sample of open items or invoices is just as valid as a projection based on account balances.

There are additional procedures which may be used to increase the response rate. A method stressed in the field of survey sampling may be of help here. Great care is devoted to the wording of an accompanying letter to emphasize the importance of the request and the need for a response, instead of using a casually phrased routine slip such as is often attached to a statement. This effort will be most productive when the recipients of the request are individuals rather than business organizations.

Responses will be further increased if a special effort is made to facilitate the process by using simple attached forms and stamped or business reply envelopes which do not require postage.

In the 1956 study of public utility retail customer confirmations mentioned above, additional random samples of 4000 requests were sent, based on negative confirmations. One group was supplied with addressed return envelopes not requiring postage while the other group was not. In evaluating the results of the test it must be remembered that replies from negatives are received only when a purported error is to be reported. The response rate for negatives without the reply envel-

[8]C. S. Warren, "Confirmation Reliability: The Evidence," *Journal of Accountancy*, February 1975, p. 87.

opes was 7.0%, while for those with reply envelopes, it was 8.3%. With samples of 4000 each, this difference was found to be statistically significant.

Other means of improving the response rate are suggested by the Canadian Institute of Chartered Accountants:

1. Provide information which the debtor can be expected to know.
2. Send communication forms to specific responsible officials of the debtor.
3. Use the client's letterhead for communication forms where possible.
4. Use confirmation information other than account balances where the debtor may experience difficulty in assembling balances for specific amounts.[9]

However, regardless of all efforts, there will be some failures to respond even after a second, and perhaps third, request. To obtain a valid and useful projection from a probability sample of confirmations, it will be necessary by one method or another to establish the validity or error of every sampling unit, for every chosen recipient, when a response is not received. Often this objective can be accomplished for those who do not respond by resorting to a variety of alternative audit procedures for receivables. Noting subsequent payment and examining sales orders, sales and shipping documents, and other evidence are appropriate steps.

However, such alternative audit procedures may not provide the same *quality* of audit evidence as a direct response from a recipient. The very existence of the need for direct confirmation in spite of the availability of these other procedures is evidence of the difference in the quality of these alternative sources of audit data.

The most widely used alternative procedure, consisting of applying cash receipts for periods subsequent to the confirmation date to the accounts receivable, has been criticized by some as being of doubtful validity.

Wright and Schoen state:

> Other accountants feel that reliance solely on cash valuation may be risky because it does not substantiate the accounts receivable balance. It is possible that payment has been made for merchandise shipped after the confirmation date or that a recorded payment is not genuine. Further, the fact that a payment has been made does not provide proof that the debtor had no claims against the company (i.e., for defective goods) at the confirmation date.[10]

[9] *Confirmation of Accounts Receivable*, p. 25.

[10] C. G. Wright and D. M. Schoen, "Receivables," in J. A. Cashin (ed.), *Handbook for Auditors*, McGraw-Hill Book Company, New York, 1971, pp. 21-1b.

Any projection from or conclusion about the validity of the accounts receivable balances, whether based on a sample or a 100% examination of the individual account balances, must be accomplished with equal validity or authority for *all* of the accounts.

The auditor must be satisfied that all individual account balances have been treated in some manner with an equal or substantially equivalent evaluation.

Lobbecke and Neter suggest that separate estimates be made for three categories of responses to receivables confirmation:

1. Those accounts which were supported as being accurately stated by either direct confirmations or by completely adequate alternative procedures.
2. Those accounts which were found to be in error by the direct confirmation with the customer and by adequate alternative procedures.
3. Those accounts where no response was received and completely adequate alternative procedures were *not* available.[11]

Of course, all three of these projections can be obtained from a probability sample.

However, it must be emphasized that if a projection of the audited value of the receivables is desired, a determination must be made, by one method or another, for *every* sampling unit in the sample, without exception. If, for any reason, no such determination is made for some of the accounts in the sample, it must be remembered that there are probably others of a similar nature in the population. A projection ignoring the missing information then does not apply to the whole population. Nevertheless, the auditor may, as a matter of policy, decide to consider those accounts for which no response was received, and for which adequate alternative data could not be obtained, as completely in error or as correct. Obviously, if a projection is made with accounts for which no determination could be made if considered to be completely in error, a very conservative estimate results. If in spite of the resulting probable over-statement of the projection, no material error results, the auditor can be assured that one does not exist in the account balances. However, if the projection indicates the possibility of a material error the auditor will now know the possible exposure.

CONFIRMATIONS AND NONSAMPLING ERRORS

The projection of the audited total value for the entire population, regardless of the sampling method used, at best provides a value equiv-

[11]Loebbecke and Neter, op. cit., p. 46.

alent to an *equal* complete coverage. In other words, the result is a projection equivalent to the results obtainable from a 100% sample where the method used to audit the individual items in the sample are applied to the entire population.

Thus, it is evident that if the units in the sample are improperly or incorrectly audited, that is, errors are claimed where there are none or not recognized when they exist, the projection provides an estimate of what would have resulted had the whole population been audited in the same erroneous fashion.

The projection of the results of a sample of confirmations is therefore dependent on the accuracy and usefulness of the responses obtained from the respondents or from alternative procedures.

If there are errors, they are referred to as nonsampling errors because such errors could occur even during a 100% audit. They are not related to the fact that a sample was used.

The validity of the projection is determined by the ability and willingness of the respondent to respond *accurately*. Responses that say the account balances are correct when they are not constitute the most serious problem. Responses that say the balances are incorrect when they are not can usually be corrected by the ensuing audit investigation but they do result in considerable wasted audit time.

While little is generally known about the accuracy of responses to confirmation requests, some experiments have disclosed interesting information in this area.

In some of these experimental studies account balances used in the confirmation sample were deliberately misstated. Where positive confirmations were used, the detection rates for the incorrect balances ranged from 26.2 to 100%.[12] In three of these experiments, negatives were also used. The rates of detection for the negatives were lower than for the corresponding positives:

	Detection Rate	
Study	Positive Confirmations, %	Negative Confirmations, %
A	54.1	39.7
B	32.1	18.3
C	26.2	16.3

[12]The results of the experiments are summarized in C. S. Warren, "Confirmation Reliability—The Evidence," *Journal of Accountancy*, February 1975, p. 85ff. Warren notes that the 100% detection rate mentioned above may be misleading in that it is based on only 13 responses.

A difference in detection rates for positive and negative confirmations was also shown by the previously cited study of public utility confirmations. Of those asked for negative confirmation 2.3% reported errors which were considered to be significant by the auditors. The corresponding rate was 5.1% for responses to the positive confirmations.

These results would seem to indicate a greater accuracy as well as a higher detection rate for positive confirmations. However, it should be remembered that the *ability* to detect errors will vary with the type of respondent. Thus, confirmations sent to retail customers will go to many recipients whose records, if any, may limit the ability to detect the errors.

In general, it is probable for most confirmations that complete detection of all errors cannot be expected even where a 100% sample is used. Most likely there will be some failures to disclose errors in account balances. Thus, the rate and extent of errors indicated by confirmation responses either from a 100% mailing *or* from a sample mailing are likely to be understated.

It should be emphasized that the auditor should recognize the fact that a 100% sample, a probability sample, or any other kind of sample, cannot achieve the determination of the "true" extent of the error. The estimate of error rates may be expected to be *lower* than the "true" rate, no matter whether based on a sample or complete coverage. The auditor must recognize the limitations of the protection provided.

GENERAL CONSIDERATIONS

Despite the fact that the confirmation procedures are an imperfect tool, it nevertheless must be recognized that direct communication with the debtor is the best of the alternatives available to establish the authenticity and correctness of the balances.

In its publication on the subject of confirmations, the Canadian Institute of Chartered Accountants recognizes these difficulties and makes the following recommendations:

1. Auditors should use the positive rather than the negative method;
2. Selection of accounts for confirmation should be representative and not influenced by the likelihood of response;
3. Alternative procedures should be used for accounts which cannot be confirmed by direct communication and by follow-up requests;
4. Traditional communication methods should be improved to achieve more replies.[13]

[13]*Confirmation of Accounts Receivable,* op. cit., pp. 1 and 2.

Chapter 6

Other Applications

ALLOWANCE FOR BAD DEBTS

Traditionally the aging of accounts receivable has been used as a device for establishing an allowance for doubtful accounts. This is accomplished by totaling the account or invoice balances for various age groups (for example, 30 to 60 days, 60 to 90 days, etc.). Generally a flat percentage established by the auditor, which may vary from age group to age group, is then applied to the aged totals.

While the same process can be used with the aid of a probability sample to replace a 100% aging, more direct and less costly methods may be used based on the statistical sample.

An allowance for bad debts can be achieved by taking a probability sample of all accounts receivable balances or invoice amounts and determining an estimate of the amount deemed to be a doubtful collection for each account in the sample. A projection of this total doubtful amount in the sample to provide a value for the whole population can then be attained.

To accomplish this, the usual approach will be to make use of a difference or ratio estimate.[1] Of course, a direct projection of the true (collectible) value of the receivables is possible, based on the book value of the account balances less any doubtful account amounts. The difference between this projection and the book record total represents the allowance for doubtful accounts. However, the direct projection method is likely to be less efficient statistically.

The method of evaluating the collectibility of each account balance is based on a judgment evaluation by the auditor for each account balance in the sample. In accomplishing this determination, the auditor may take into account various factors including the age of the account *and* other considerations.

[1]See H. Arkin, *Handbook of Sampling for Auditing and Accounting*, 2d ed., McGraw-Hill Book Company New York 1974, chap. 12, for a discussion of these methods.

In many respects this method is more valid than applying a fixed percentage to the total value of each age group. When used, the method provides an opportunity to carefully evaluate each sample account balance, thus permitting the application of other considerations beyond mere age. The resulting projection provides the same result as that which would have been obtained if *all* accounts receivable balances had received the same kind of careful evaluation.

This approach replaces the time-consuming and costly detailed aging of all accounts and at the same time produces a more valid estimate for the entire population. Further, experience has shown that when accounts are numerous, errors will be made in a 100% aging.

However, if it is desired to use the more customary method, estimates can be made for each age group from a statistical sample and the percentages applied as usual. Again the method eliminates the need for 100% aging. In the simplest situation, when this latter method is used, the auditor may decide to apply a fixed percentage to all account amounts due longer than some fixed period, say 60 days. The probability sample may then be used to provide a projection of the total dollar value of the account balances in that age category to which the percentage can then be applied.

If the sampling error for any of the above estimates projected from the statistical sample is reasonably small, the point estimate obtained from the sample may be used. If the sampling error is large, the auditor will have to select either the upper confidence limit for the most conservative estimate (largest allowance) or use the lower confidence limit if a minimum allowance is found to be appropriate.

INTERAIRLINE BILLINGS

In airline travel, tickets containing flight coupons are often sold by one airline but picked up by another airline, when their planes are used. Many of the major international airlines have agreed to the use of sampling for accounting for interline passenger coupons.[2]

The International Air Transport Association (IATA) manual notes that due to increasing interline transactions and the ever more complicated fare structure a method was sought to limit the inevitable increases in skilled staff.[3] The manual notes that "as a result of consid-

[2]The following description of the method is based on *IATA Revenue Accounting Manual*, International Air Transport Association, April 1, 1974, (revised January 25, 1977), Geneva, Switzerland.

[3]Ibid., chap. B1, p. 1.

erable research, it has been found that passenger interline accounting can be carried out to a degree of accuracy that can be controlled and is acceptable, by the use of the scientific principles of random sampling."

Under this system a provisional invoice is submitted by each airline to the airline which sold the flight coupons picked up. This provisional invoice is essentially a total of the value of the flight coupons for the universe.

The universe is defined as consisting of all coupons, regardless of value, which have been accepted for travel, refund, or exchange, with certain specified exceptions.

The provisional invoice amount is then adjusted by the results of the examination of a sample. The sample is selected on the basis of a digit selected at random and issued monthly by the manager of the Clearing House.

All flight coupons with numbers ending in that digit are then included in the sample. A sample of at least 60 is required with a 100% examination required for universes of 75 or less. Methods are provided to cover the situation where insufficient flight coupons are selected either by reference to the next to last digit or to the next higher digit. Each flight coupon is then evaluated at its exact value including appropriate commission rates.

A sampling constant is computed by dividing the total of the universe as reflected in the provisional invoice by the total of the sample flight coupons evaluated on the basis used in the provisional invoice.

This ratio, which of course will always be greater than 1, is used as an expansion multiplier and applied to the total of the exact values of the sample flight coupons.

The difference between the total on the provisional invoice and the projected sample results then represents the adjustments to that provisional invoice.

TAXES AND SAMPLING

There are two major areas where statistical sampling is applied to tax data:

1. The compilation by sample estimate of certain types of data for use in conjunction with business tax returns allowed by the tax services
2. The audit of individual business tax data upon which tax returns are based

TAXPAYER APPLICATIONS

The U.S. Internal Revenue Service (IRS) has permitted the use of sample projections in compiling data to be used for tax return purposes. Not all of these permitted applications are known to the author, since some have been approved by letter and not published in tax regulations.

Some examples of such applications include:

1. Establishing current sales in a revolving credit plan
2. Establishing a reserve for unredeemed trading stamps
3. Making LIFO evaluations
4. Establishing a reserve for unexpired subscriptions for a large circulation magazine

The first three of the above applications are permitted by published tax procedures. The other application is known to have been permitted.

At present, there is no published tax regulation or procedure covering the general application by the taxpayer of probability sampling for compiling such data for use in a tax return, though individuals have been granted permission to use these methods in various situations.

The increasing complexity of the data required of the taxpayer has resulted in increasing use of probability samples by taxpayers and increasing acceptance by the IRS of such sample data as a basis for tax return reporting.

REVOLVING CREDIT PLANS

When a revolving credit plan is used by a vendor for tax purposes, the vendor may report the gross profit on a portion of these amounts on the installment method of reporting income. The profit component of each installment payment is included in income when that payment is received, spreading the tax over the payment period rather than the sale period.

To establish the amounts involved, it is necessary to analyze each account and fix the portion eligible. When the accounts are numerous, a 100% determination accomplished by examining all accounts is an extended and costly task.

The IRS permits the use of statistical sampling for this purpose and prescribes guidelines for such usage in Revenue Procedures 64-4 and 65-5.

The Revenue Procedure provides that a sample may be used for this purpose if one of three methods is selected.

1. The taxpayer may use any sampling procedures that are in accordance with generally accepted probability sampling techniques.
2. The taxpayer may elect the procedures specified in Section 5 of the Revenue Procedure titled "An Acceptable Procedure for Sampling."
3. The taxpayer may elect to use a 100% sample.[4]

If a sampling technique is used under option 1, above, "the procedure must be documented and made available at the request of the District Director of Internal Revenue. . . ."[5] The IRS also requires that the records and the procedure used be preserved for future use to make it possible to verify the sample selection methods and the computations performed, and to verify that the frame adequately covers the population.

The sample size which the IRS requires is established by Revenue Procedure 64-4, which says "the sampling procedure and sample size must be such as to provide for a relative sampling variability in the amount of gross profit to be deferred (i.e., the coefficient of variation of the estimate expressed as a percent) of no more than two percent."[6]

The coefficient of variation specified above is calculated by dividing the standard error of the mean (or total) by the point estimate of the arithmetic mean (or total) where the standard error of the average for an unrestricted random sample is

$$\sigma_{\bar{x}} = \frac{\sigma}{\sqrt{n}} \sqrt{1 - \frac{n}{N}}$$

The coefficient of variation then is

$$CV_{\bar{x}} = \frac{\sigma_{\bar{x}}}{\overline{\overline{X}}}$$

This method of specifying the sample size bypasses the problem of specifying a confidence level. The resulting sample size is fixed regardless of the confidence level selected, but the sample precision attained will vary with the confidence level chosen.

[4]IRS Revenue Procedure 64-4.

[5]Ibid. Specifically with reference to sample design, there must be submitted with the tax return the following:

(a) Description of the frame or frames
(b) Definition of the sampling unit or units
(c) The procedure for drawing the sampling units
(d) Computation on which the decision of the sample size was based
(e) The formula or procedure for calculating the relative sampling variability in the estimate of the gross profit to be deferred as well as its computation.

[6]Ibid.

The resulting sampling error (sample precision) at various confidence levels for a coefficient of variation is shown below.

Confidence Level, %	Sample Precision as Percent of Estimate
90	3.29
95	3.92
99	5.16

When an unrestricted random sample is used, where the standard deviation is large compared to the arithmetic mean, very large samples will be required. Unfortunately, with the greatly skewed populations of accounting values, the standard deviation is often several times the value of the arithmetic mean. For instance, for the distribution of invoices in Table 3-3 the standard deviation of the values, $712.64, is almost three times the value of the arithmentic mean, $255.26. For the accounts receivable balances (Table 3-4), the standard deviation is more than twice the value of the arithmetic mean.

For a very large population, the sample sizes required for an unrestricted random sample to achieve the required coefficient of variation of 2% of the value of the arithmetic mean (or total), where the standard deviation is one of several multiples of the arithmetic mean, are shown below.[7]

Ratio of Standard Deviation to Arithmetic Mean	Required Sample Size
1.0	2500
1.5	3750
2.0	5000
2.5	6250
3.0	7500

These very large sample size requirements generally will dictate the use of sampling methods which are more efficient than unrestricted random sampling, such as stratification, ratio estimates, etc., to reduce the sample size to a more reasonable basis. However, the remaining

[7]The required sample size for a large population can be computed from the following formula when the value of the coefficient of variation of the arithmetic mean is specified:

$$n = \frac{\sigma^2}{[(CV_{\bar{x}})\bar{X}]^2}$$

sample size, even when more efficient techniques are used, is likely to be quite large. Nevertheless, because of the very large volume of this type of account the use of a probability sample is likely to be much less costly than a 100% evaluation.

RESERVE FOR UNREDEEMED TRADING STAMPS

IRS Regulation 72-36 provides that when a taxpayer uses an estimate of the value of future redemptions of trading stamps for tax purposes, the estimate may be achieved using probability sampling. The requirements for such usage of probability sampling are detailed in Revenue Procedure 72-36.

For this purpose a sample of redeemed trading stamps is taken to estimate the percentage of stamps issued in a given year which were redeemed in the same year and in each succeeding year. This computation is accomplished by taking a probability sample of stamps redeemed during the tax year and determining in what year each was issued.

The Revenue Procedure provides that "the taxpayer may use any sampling procedures that are in accordance with generally accepted probability sampling techniques."[8] However, the illustration provided and many of the instructions given are based on the use of a "replicated sample."[9] Any probability sampling technique used is subject to review by the District Director of Internal Revenue.

The Revenue Procedure further requires that the design of the study be the responsibility of a competent sampling expert with a cautionary note to the effect that probability sampling is a highly specialized field and that otherwise competent statisticians may not be qualified in the field of probability sampling. No specific qualifications are given for such a "competent sampling expert" but the IRS requires that the expert's qualifications be available to the District Director of Internal Revenue.

The Revenue Procedure requires that the sample size used be 1 stamp per 1.5 million issued during the previous year with a minimum of 1000 and a maximum of 50,000. These sample size requirements may later be reduced to 1 per 3 million with a minimum of 1000 and a maximum of 40,000 if certain tests are met. Depending on the number of years in which the tests are met, a further reduction ranging from a

[8]IRS Revenue Procedure 72-36, Section 5.02.

[9]For an explanation of replicated sampling see H. Arkin, op. cit. chap. 14.

requirement of 1 per 5 million to 1 per 10 million with a minimum of 1000 and a maximum of up to 30,000 may be allowed.

The number of stamps to be redeemed in future years is based on a point estimate obtained by applying the inverse of the sample ratio for each issue year over a period of 10 years. These estimates are then divided by the number of stamps actually issued each year to obtain a redemption-rate percentage for each "lag" year. These rates are applied to each "lag" year to establish the number which will be used in future years for an initial redemption pattern to "reconstruct" future redemptions and to use in various other methods of establishing mathematical models.[10]

LIFO APPLICATIONS

IRS regulations permit the use of the LIFO method of inventory valuation. In a period of inflation with rapidly increasing prices the use of such a method sharply reduces the amount of income taxes to be paid by a business when a material inventory is involved.

IRS Regulation 1.472-1 states that "any taxpayer permitted or required to take inventories ... may elect with respect to those goods ... to compute his opening and closing inventories ... under this last-in–first-out (LIFO) inventory method."

One of the methods which may be applied is the "dollar value pool" method. Under this technique "pools" of homogeneous items similar to the strata of stratified sampling are created. Ordinarily the double extension method is used under which the year-end inventory is priced at both "base period" (initial) prices and current period prices. When there is an extensive variety of items in the inventory or there are other complexities, an index may be computed based on a sample of stock items rather than double extending all items.

The appropriateness and reliability of such an index based on a sample is subject to review by the IRS and the method of computation, in detail, must be filed with the return.[11]

[10]A more extensive treatment of the method of creating such models is covered in H. J. Davidson, J. Neter, and A. S. Petran, "Estimating the Liability for Unredeemed Stamps," *Journal of Accounting Research*, 1967, pp. 186–207.

[11]IRS Regulation 1.472-8(e) states that "a taxpayer using either an index or link chain method shall attach to [the] return for the first taxable year ... a statement describing the particular ... method used in computing the index. Adequate records must be maintained by the taxpayer to support the appropriateness, accuracy and reliability of an index. . . ."

This index is essentially the ratio between the ending inventory quantities at current year cost and the ending inventory quantities at base year cost for the stock items in the sample.

$$\text{Index} = \frac{\text{current year's cost of ending inventory}}{\text{base year's cost of ending inventory}}$$

As prescribed in IRS Regulation 1.472-8, this index is then applied to the *increase* between the *current* year inventory (ending inventory) quantities at base-year costs and the *base*-year quantities at base-year costs. This increase is called the "layer." The index is applied to the "layer" to establish the excess of FIFO over LIFO. This computation is performed only if the layer represents an increase. When the layer results in a decrease, a new LIFO is not computed but results in the reduction of the most recent layer.

The computation is accomplished for each year thereafter with the LIFO ending inventory for each year becoming that inventory value at base-year costs plus the "layer" value multiplied by the LIFO index for that and prior years. The excess of FIFO over LIFO is charged as an increase in the cost of goods sold for the current year.

A number of taxpayers have used statistical sampling for this purpose. It is to be recognized that the projection of a double extended sample is a ratio. The ratio is computed by dividing the total inventory value in the sample by the total inventory value at base period prices. The sampling error of this ratio must be computed by ratio estimate methods.

The basic formula for obtaining a point estimate of a ratio from a sample for LIFO index purposes is

$$\text{LIFO index} = \frac{\text{total sample value at current prices}}{\text{total sample value at base period prices}} = \frac{\Sigma y}{\Sigma x}$$

The basic formula for the sampling error of the ratio for LIFO index purposes is

$$SE_r = \frac{t\dfrac{\sigma'}{\sqrt{n}}\sqrt{1-\dfrac{n}{N}}}{\overline{X}}$$

where SE_r = sampling error of estimated ratio

t = factor reflecting selected confidence level (e.g. 1.96 for 95% confidence level)

$$\sigma' = \sqrt{\frac{\Sigma(y - rx)^2}{n - 1}}$$

$$= \sqrt{\frac{\sigma_y^2 + r\sigma_x^2 - 2r\sigma_{xy}}{n-1}}$$

$$\sigma_{xy} = \frac{\Sigma xy}{n} - \overline{xy}$$

x = value of each stock item in sample at base period prices
y = value of each stock item in inventory at current period prices
$\overline{x}$ = arithmetic mean of x values in sample
$\overline{y}$ = arithmetic mean of y values in sample

It may be expected that when an index is based on a sample, the IRS will insist on a relatively low sampling error.

In this respect, it must be remembered that the operating portion of the index is obtained by subtracting 1 from the index to obtain the *change*.

In a period of rising prices this index will exceed 1. For instance, the index might be 1.12; subtracting 1 from the value yields .12 or the relative change. It may be expected that the IRS will insist that the computed sampling error be relatively small compared to the change factor

$$\frac{SE_r}{(\text{Point estimate of } r) - 1}$$

This can be accomplished only by a large enough sample, an efficient sample design, or a combination of both.

INTERNAL REVENUE SERVICE AUDIT APPLICATIONS

Confronted with huge volumes of entries, transactions, and documentation in large corporations, the IRS recently has been training its personnel in the use of statistical sampling for audits of business tax returns.

An article in *The New York Times* dated March 13, 1979, calls attention to the fact that while "for years the Internal Revenue Service has used statistical sampling techniques to select tax returns for audit . . . for the last few months, however, it has been quietly expanding the technique and applying it to the actual tax audits of the country's biggest corporations. In fact, the use of these techniques in auditing business tax returns is being extended rapidly with resulting increased collection of business taxes."[12]

[12]"New Approach on Tax Audits," *The New York Times*, Mar. 13, 1979, p. D2.

In the Peat, Marwick, Mitchell & Co. *Executive Newsletter* it is noted that "only a few taxpayers have been affected so far, but the IRS apparently plans to extend the use of statistical sampling broadly over the next few years."[13]

It must be recognized that the more specific and limited objectives of an IRS sampling of corporate records will result in somewhat different views and approaches from those of the public accounting firm.

Here the purpose is to establish the reasonableness of individual values submitted on tax returns. Since much of the data so presented must meet IRS requirements and are not relevant to the objectives of the public accounting firm, the IRS cannot completely rely on the fact that the corporation's financial statement was audited by a reputable public accounting firm.

The public accounting firm's examination may not be concerned with values which do not appear directly on the financial statements, such as a listing of investment tax credits. In fact, their views may even be different with reference to the same values. The public accounting firm is generally interested in the assurance that assets are not overstated. The concern of the IRS is that they are not understated. Some techniques used by public accounting firms are directed solely at such overstatements (see Chapter 7).

Since it is not uncommon for individual deductions claimed by corporations to be disallowed by the IRS, the objective of the IRS is either to estimate the amount of such disallowed deductions or make adjustments to submitted data.

Typical areas to which statistical sampling methods are applied include disallowances in travel and entertainment expenses, in expenses that should have been capitalized, disallowed investment tax credits, etc.

An IRS Manual Supplement (42 G-386) dated September 27, 1978, discloses the method of treatment of some of the special problems confronted when statistical sampling is used in audits of business tax returns.

The projection of disallowances or adjustments to the population from the examination of a limited number of sampling units will, of course, produce a point estimate which is subject to a sampling error. If the sample is large enough the sampling error may be considered innocuous and be ignored. However, with the limited audit resources and time available in this, as in almost all other audit situations, the

[13] *Executive Newsletter*, Peat, Marwick, Mitchell & Co., vol. V, no. 2, Feb. 26, 1979, p. 3.

sample size is not likely to be of such a magnitude and the sampling error is likely to be significant.

To overcome this problem, the IRS manual indicates that the *lower* limit of the resulting confidence interval estimate will be used as a basis for claiming an adjustment to tax return data. For this purpose, a *one-sided* 95% confidence level is prescribed. The IRS manual notes that "as a general rule, the proposed population adjustment will be determined, such that 95% of the time, it will not be greater than the actual adjustment obtainable by a 100% examination of the population."[14]

The manual then continues by noting that "the above result will be attained by using the lower limit of the estimated population adjustment at the 95% confidence level. . . . This lower limit will generally be computed by subtracting the sampling error . . . from the population adjustment."

This approach means that the IRS cannot produce a precise projection from the sample and is willing to concede the amount of the sampling error to the taxpayer. This approach means that the smaller the audit sample size the bigger the sampling error and hence the greater concession to the taxpayer.

The *Executive Newsletter* of Peat, Marwick, Mitchell & Co. notes that "the IRS policy is not to propose an adjustment exactly equal to the sample findings. Instead, it would propose a smaller adjustment, which naturally becomes favorable to the taxpayer in the case of a disallowance. Under the IRS approach, the odds are 19 to 1 against overassessing a deficiency."[15]

Under the procedures given in the IRS manual the protection may be against or in favor of the taxpayer. However, if the point estimate calls for an adjustment against the taxpayer but the lower limit is negative (in favor of the taxpayer), due to a small sample, the sample projection is inconclusive. The sample will either be extended or the sample projection dropped in favor of the individual disallowances found in the sample.

In similar manner, if the point estimate favors the taxpayer (is negative) and the lower limit favors the government (is positive), again, because the result is inconclusive, the sample will either be extended or the projection dropped in favor of the individual items in the sample.

Thus, in general, when the sampling error is greater than the point estimate, no matter its direction, the sample will be extended or the projection dropped since it is inconclusive.

[14]IRS Manual Supplement 42G–326, Sept. 27, 1978, p. 2.

[15]*Executive Newsletter*, Peat, Marwick, Mitchell & Co., vol. V, no. 2, Feb. 26, 1979.

The IRS sampling procedures may be applied manually to noncomputerized data or through a computer program (the PAL system) especially devised by the IRS to be applied to computerized records.

An extensive training program for IRS examination personnel in the use of statistical (probability) sampling in this situation has been developed, and it may be expected that many more applications for this technique will be found.

The use of a statistical sample projection instead of prior judgment sampling methods will encourage compliance by business taxpayers and, by achieving a result similar to 100% examination, result in large additional tax payments where taxpayers do not comply.

Chapter 7

Dollar Unit Sampling

A primary concern of the independent public accountant during the audit of the financial statements of a commercial enterprise is the possibility of a material overstatement of the assets of the company.

While the substantive test (variables estimation sampling) is designed to estimate the total value of any existing overstatements, under certain unusual circumstances it may fail to do so.

When the usual probability sampling methods are used, the sample is selected by drawing an unrestricted random sample of sampling units, such as invoices, vouchers, checks, or other physical evidence of accounting transactions. From this point on, such documentation will be referred to as *physical sampling units.*

Such methods of selection of a sample are, of course, statistically valid and widely used. There is the possibility, however, that there may be a few sampling units among many which in themselves contain a material error. If so, the probability that any one of these erroneous sampling units will be included in the sample is small. They would, therefore, probably not be detected nor would they be reflected in the sample projection. Thus, it is possible in such circumstances to fail to detect a material error.

To protect themselves against such a possibility, the auditors usually examine all transactions large enough to individually contain a material error as well as sampling the remaining transactions. This results in a stratified sample.

However, this does not provide full protection against the possibility that there are a *few* such transactions among many which, while not large enough to contain a material error individually, may collectively add up to such an error.

For instance, assume that there are 10 sampling units, each containing an error amounting to a little more than one-tenth of the material amount, among 10,000 transactions. The probability that *any* of these sampling units would be included in a sample of 200 is only 9.6% or about 1 chance in 10. Thus, such a material error might well go undetected.

In general, this risk can be reduced by stratification, especially if a larger proportion of the units in the higher-value strata are sampled. However, stratification does not completely eliminate the problem. Based on the principle that the higher-value items are of greater concern to the auditor, a new method of sampling which provides a probability of including a sampling unit in the sample proportional to its dollar value has been suggested.[1] This new method is called *Dollar Unit sampling.*[2]

PRINCIPLES OF DOLLAR UNIT SAMPLING

This new method involves a change in the sample selection technique. The sampling unit involved here is not the *physical* sampling unit previously used, but rather is defined as the *dollar unit,* or each of the individual dollars which make up the total dollar value of the account being sampled.

Thus, if vouchers are being sampled, the population consists of the total dollar value (number of dollar units) of all of these vouchers and the sampling unit is *each* individual dollar of that total.

The sample selection method involves an unrestricted random sample of these dollar units (dollar bills), *not* the vouchers.

The simplest of the sample selection methods in this technique is a systematic sampling of the dollar units. The sample selection techniques will be described in detail later, but the result is an equal probability that any single dollar unit (individual dollar) will be included in the sample regardless of the voucher with which it is associated.

This can best be envisioned by conceiving that a pile of dollar bills is created by cumulating the values of each of the vouchers in the form of single dollar bills. The entire pile would then compose the population. Dollar bills would then be selected randomly from this pile and the *physical* sampling units which include the selected dollar bills audited.

Thus, if a group of 1000 vouchers (physical sampling units) with a book record value of $100,000 is to be tested, a pile of 100,000 dollar bills is envisioned.

Assume that some of these vouchers are in fact overstated. This would be equivalent to observing that some of the dollar bills in the

[1]Those with a background in sampling theory will recognize this method as sampling with probability proportional to size (PPS).

[2]This method of sampling also is referred to as *monetary unit sampling.*

population are fictitious or counterfeit. If the objective is to establish the amount of the *overstatement*, the sampling problem can be reduced to establishing the percentage of the dollar bills in the population which are counterfeit (overstated). This is an attributes sampling situation and the methods developed for the appraisal of attributes samples may be used.

This method of sample selection overcomes the difficulty that caused this technique to be developed, since the concentration of overstated dollars in a few physical sampling units is not now relevant. The physical sampling units are not sampled. The probability that a fictitious dollar unit resulting from the overstatement of the value of a physical sampling unit will be included in the sample is strictly proportional to the percentage of overstatement.

Thus, the probability that a given *physical* sampling unit will be included in the sample becomes proportional to its value. The more a physical sampling unit is overstated, the more likely it is to be included in the sample, since its dollar value (number of dollar units) would be increased by the overstatement.

The selection of a particular dollar unit identifies the *physical* sampling unit with which it is associated. This physical sampling unit is then audited to establish the magnitude of any overstatement. A variety of approaches are available to project the result to the entire population and determine the interval estimate of the overstatement. These methods are discussed later.

SAMPLE SELECTION METHODS

The selection of the dollar units to be included in the sample can be accomplished randomly by using either systematic or random number selection techniques. Often the sample selection can be most conveniently accomplished by systematic selection.[3]

When the systematic selection method is used, the dollar values of the physical units (i.e., vouchers, etc.) must be added cumulatively.

For instance, assume a series of inventory stock items with a reported total record value of $360,000. The individual item values are added cumulatively as shown in Table 7-1.

The sample size can be determined by a method described later. Assume that a sample of 300 dollar units is desired. Since there are a total of 360,000 dollar units in the population, the sampling interval for

[3]Since dollar units are selected, it may be assumed that there is no fixed pattern or trend in the sampled data (overstated dollar units). This fact overcomes the general objections to the use of systematic sampling.

Table 7-1

Stock Item Number	Dollar Value	Cumulative Dollar Value	Dollar Unit Numbers
1146	$ 200	$ 200	1- 200
1147	500	700	201- 700
1148	50	750	701- 750
1149	600	1,350	751-1350
1150	100	1,450	1351-1450
1151	40	1,490	1451-1490
1152	50	1,540	1491-1540
1153	300	1,840	1541-1840
*	*	*	*
*	*	*	*
*	*	*	*
		$360,000	
	$360,000		

a systematic sample can be determined by dividing the population size by the desired sample size.

$$\frac{N}{n} = \frac{\$360{,}000}{300} = 1200$$

Every 1200th dollar sampling unit is selected, using a random start. Thus, if a random start at dollar unit 150 is established, the first dollar unit to be included in the sample would be #150, the second selected dollar unit is #1350, the third is #2550, etc. These dollar units are then located within the physical sampling units with which they are associated. Thus, in the example above, the first selected dollar unit (the 150th) is included within stock item #1146, the second (the 1350th) within stock item #1149, etc. The associated physical units (in this case #1146, #1149, etc.) are audited to establish any overstatement in these physical sampling units.

Another selection method which may be used when the values of the physical sampling units are reasonably uniform is the equivalent of the page and line number selection method.[4] This method makes the cumulation of the values of the physical sampling units in the population unnecessary.

Assume that there are a total of 5000 physical sampling units *with no item in excess of $500*. Four-digit random numbers are obtained from

[4]See H. Arkin, *Handbook of Sampling for Auditing and Accounting*, 2d ed., McGraw-Hill Book Company, New York, p. 38, for a detailed discussion of this method.

a random number table. The random numbers between 1 and 5000 so obtained designate the physical sampling units. Three-digit random numbers between 1 and 500 then indicate the dollar unit within the physical unit. For instance if the four-digit number is 0459 and the three-digit number is 152, the dollar sampling unit selected is the 152nd on physical sampling unit #459.

If the physical sampling unit involved has a value *less* than the selected three-digit number, the dollar unit is nonexistent and *both* the four-digit physical unit number and the three-digit number are discarded and a new pair drawn. This process is continued until the desired sample size is attained.

APPRAISAL METHODS

Once the sample has been drawn and the associated physical units audited, the sample results must be projected to obtain the total overstatement for the entire population. Several methods have been suggested for this purpose.

It is important to note, at this point, that the above discussion of Dollar Unit sampling has dealt with this technique with respect to estimating overstatement only. Some of the evaluation methods described below attempt to take into consideration offsetting understatements. The effectiveness of each of these approaches is discussed below after the description of each of the methods.

The Attributes Approach

The attributes approach is limited to establishing the gross overstatement, if any, without any consideration of possible offsetting understated values. It will be seen later that the attempt to allow for such understatement in establishing a net overstatement is of questionable validity for the methods described below.

If the total value of the physical sampling units is thought of as a pile of dollar bills, some of which are bogus (representing the overstated portion), it is possible to estimate the percentage of the dollar units which are bogus by using estimation sampling for attributes methods.

In the above example, the first dollar sampling unit selected for inclusion in the sample was #150. It was associated with stock item #1146. This stock item value was audited and it was established that it was overstated by $75 (75 dollar units).

The next step is to establish whether the selected dollar unit (#150) is one of the overstated (bogus) dollars. This can be accomplished by resorting to a fixed decision rule established in advance. For instance, it can be decided that if a physical sampling unit is overstated, the *last* dollar units within that physical sampling unit, in a number equivalent to the overstatement, will be the ones considered bogus.

If the number of the randomly selected *dollar unit* is one of those considered to be bogus by this rule, it would be considered to be an overstated dollar unit. The number of such overstated (bogus) dollar units in the sample is counted and divided by the sample size in dollar units to obtain the point estimate of the percentage of overstated dollar units in the population. Multiplying the total book record value by that percentage results in a point estimate of the total value of the overstatement.

An interval estimate may be obtained by applying the usual attributes estimation sampling evaluation method. For instance, assume that the audit is accomplished for the sample used above. The first dollar unit to be included in the sample, when the systematic selection method was used, is #150, which was associated with stock item (physical unit) #1146. The book record value of stock item #1146 was $200, but the audit disclosed it to be overstated by $75. Thus, using the decision rule, the last 75 dollar units of the 200 associated with stock item #1146 are considered to be bogus (overstated). The selected dollar unit (#150) is established to be one of the bogus dollar units since it is one of the last 75.

This determination results in a count of one overstated dollar unit included in the sample. If the selected dollar unit had been #50 (instead of #150), on the basis of the decision rule it would *not* be considered to be an overstated dollar unit.

The audit process is continued for all 300 dollar units in the sample. Depending on the sampling selection interval used in the systematic selection, there may be more than one dollar unit selected for a physical sampling unit (stock item). If so, the physical sampling unit is audited only once, but a separate determination is made for each dollar unit selected, depending on their positions within the physical unit. Some, all, or none may be overstated dollar units.

In that example, which involved 5000 *physical sampling units* with a total book record value of $360,000, a sample of 300 *dollar sampling units* was obtained. Assume that the audit determined that 15 of the 300 sampled dollar units were overstated. This makes up 5% of the 300 dollar units in the sample and thus provides a point estimate that 5% of the dollar units in the population are overstated dollars. Hence, the point estimate of the total overstatement of the book record value is $360,000 multiplied by 5%, or $18,000.

Of course, to be useful, an interval estimate of the total overstatement must be provided. This estimate can be obtained by resorting to appropriate tables of attributes confidence limits.[5] Since the rate of occurrence in the sample was 5%, reference to such tables provides a confidence interval estimate of 2.8 to 8.1% at the 95% confidence level. When applied to the book record value of $360,000, the interval estimate of the total dollar value of overstatement becomes $10,080 to $29,160.

An unusual feature of the approach is that it provides an ability to establish a *maximum* amount of overstatement (at a given confidence level), even when *no* overstated dollar units are found in the sample.

In the above example, if no overstated dollar units had been found in the sample, consulting tables of confidence limits based on the binomial distribution would provide an upper limit.[6] Thus, the table for zero rate of occurrence in the sample, for a sample of 300, indicates an upper limit of 1% at the 95% confidence level. Thus, there is a 95% assurance that the total overstatement, if any, would not exceed $3600 (1% of $360,000).

The confidence limits obtained by the above methods may then be interpreted in the same manner as that for an estimation sample for variables, as discussed earlier in Chapter 3.

If the *upper limit* is less than that amount considered to be material, there is a high probability (determined by the confidence level used) that *no* material overstatement exists. If the *lower limit* is more than that amount considered to be material, there is a high probability that a material overstatement exists. If the amount considered to constitute a material error falls between the two limits, there is no evidence that a material overstatement exists, though there is a possibility that it does.

In the last circumstance, if it is decided to extend the sample to resolve the dilemma, it must be remembered that attributes estimation sampling techniques were used, and the two-step sampling approach described in Chapter 2 must be applied.

Sample Size Estimation for Dollar Unit Sampling

Before conducting an audit test, it is necessary to estimate the sample size required. The method for this form of Dollar Unit sampling is similar to that for attributes estimation sampling. A minimum amount

[5]Such tables can be found in H. Arkin, op. cit., appendix F.

[6]The binomial distribution may be used in place of the hypergeometric (finite populations) since in all cases the Dollar Unit sample will be very small compared to the population size. The confidence limits for a zero rate of occurrence in the sample may be found in Arkin, op. cit., table F27.

which would be considered to be material must be established. When expressed as a percentage of the total book record value, the percentage may be construed to be similar to the MTER.

Basically, the sample has to be large enough to make possible a determination that there is or is not a material overstatement. If the sample is too small, it will merely indicate that no conclusion can be reached because the amount considered material falls between the upper and lower confidence limits.

A possible method would be to use a sample large enough to at least provide assurance that if no errors are found in the sample, there is no significant probability that a material error exists.

For instance, in the previous example with a book record value of \$360,000 perhaps \$20,000 might be the minimum amount that would be considered material. This material amount constitutes 5.6% of the book record value. By reference to Appendix A (page 173), it will be seen that the sample size required to provide an upper limit of 5.6%, if no errors are found in the sample, is between 50 and 60 at the 95% confidence level.

However, this would be a *minimum* sample size. If overstated dollar units are found in the sample, it is not likely that the resulting confidence interval would provide an ability to establish firmly a material overstatement, if it exists, as compared with a less than material amount. For this reason, a sample larger than the minimum would be needed.

The sample size to be used may then be estimated by the usual methods for attributes samples where the maximum expected rate (MER)[7] is based on a conservative estimate of the percentage of overstated dollar units in the population.

Another approach which may be effective is to use the results obtained from the first sample, if any overstatements are found, to estimate the sample size requirement for additional sampling. This is a *two-step* sampling procedure, as described in Chapter 2. The tables provided in Appendix A may be used for this purpose.

If overstated dollar units are detected, the sample used must be sufficient to indicate the existence of a material error, if it exists, by providing a *lower* confidence limit which is at least equal to the minimum amount considered to be material.

For instance, in the above example where \$20,000 or 5.6% of the book record value was considered to be the minimum amount to be regarded

[7]The MER then is used as p in the formula

$$n = \frac{p(1 - p)}{(SE/t)^2}$$

as material, an opportunity to demonstrate that such a material error exists, when it does, must be provided. The evidence provided by the initial sample may be used to assure such a capability.

For example, assume that an initial sample of 60 disclosed six overstated dollar units. This is 10% of the book record value and provides a point *estimate* of the overstatement in the population of $36,000. However, this point estimate does *not* prove that a material overstatement exists. Another sample might well produce a much lower overstatement rate and perhaps a nonmaterial overstatement amount, especially with an original sample as small as 60. In fact, another sample of 60 might well produce a point estimate of anything between 3.7 and 18.5% (at the 95% confidence level). To demonstrate the existence of a material error, it is necessary that the *lower* confidence limit of the interval estimate be above the minimum amount considered to be material. Otherwise a material error might be claimed where none existed.

While the point estimate for an increased sample will probably change, for the purpose of sample size estimation it may be used in the calculation. In the above example, it becomes necessary to find a sample size which would yield a lower confidence interval limit of $20,000 or 5.6% if the same proportion (10%) of overstated dollar units are found in the enlarged sample. Reference to Appendix A (page 197) indicates that if the point estimate of the rate of occurrence of 10% is found in the enlarged sample, and if an additional sample of 120 is used, a lower confidence limit of more than 5.6% (95% confidence level) will result.

This additional sample size requirement was obtained from Table A-2 for an initial sample size (n) of 60, a 95% confidence level, and a population size (N) of 10,000 (the largest population size in the table). Applying the 10% point estimate rate to the final sample size ($n_1 + n_2$), with the initial sample size of 60 and a second sample size of 60, this computation yields 12. For $n_2 = 120$, the result is 18, while for the second sample size of 180, it is 24. Reference to the table shows that the lower limit for $n_1 = 60$ and $n_2 = 60$ with 12 in the sample is 5.3% while for $n_1 = 60$ and $n_2 = 120$ with 18 in the sample, the lower limit is 6.1%. This last meets the requirement that the lower limit be in excess of 5.6% and thus an additional sample of 120 would be appropriate. It is to be emphasized that such a sample size determination is merely an estimate.

If the ultimate sample of 180, when audited, contains 17 or more overstated dollar units, it may be concluded that there is a material overstatement. If fewer are found, no firm determination can be made. The auditor might then conclude that there is a possibility of a marginal situation. However, Table A-2 (page 179) will provide further information which may be of help. The following figures, extracted from that table, apply to this example:

Number of Overstated Units in Sample of 180	Percent Overstatement Confidence Limits (95% Confidence Level)	
	Lower	Upper
10	3.1	9.2
14	4.8	11.9

Note that the sample consists of an initial sample of 60 plus an additional sample of 120. Six overstated dollar units were found in the first sample of 60. The minimum material error is 5.6% of the book record value.

Limitations and Advantages of Dollar Unit Sampling (Attributes Evaluation)

All forms of Dollar Unit sampling are subject to certain limitations. Some of the advantages, however, are unique to the attributes approach described above. Some additional limitations or disadvantages apply to the other Dollar Unit sampling plans described later.

It must be emphasized that while the general principle of Dollar Unit sampling does resolve a major audit problem dealing with the occurrence of rare errors which are collectively material, the technique is subject to important limitations.

Limitations

1. *The Dollar Unit method may be applied only to* overstatements.

   The method uses an approach based on the concept of fictitious dollars representing overstatements. It was noted that the greater the overstatement, the more fictitious dollars in the population and the greater the probability of inclusion in the sample.

   For understatements, the dollar units are *missing* from the population and cannot be sampled. Thus, the greater the understatement, the smaller the number of dollars for a physical sampling unit and the less likely its inclusion for audit. It is not possible to sample units which are not there. Thus, it is not possible to properly estimate understatements or *net* errors.

2. *The selection of the sample on a manual basis is awkward.*

   The method of sample selection using systematic sampling requires the cumulative totaling of all of the physical sampling units in the population. For large populations this may be impractical unless the data are computerized.

3. *The method does not permit the inclusion of physical units with zero or negative values in the sample.*

   Since the probability of inclusion of a given physical sampling unit in the sample is proportionate to the number of dollar units associated with it, there is no probability that any of these units would be included in the sample. Yet these physical units might be overstated since a zero value might properly be a negative value (i.e., a credit) or a negative figure might be more negative.

   If there is a *possibility* that negative values could occur in the population, the method cannot be used unless there is a separate sampling of these zero or negative values using a traditional sampling plan.

4. *The method cannot be used unless individual and total book values are available and hence cannot provide estimates of unknown population values.*

Advantages

1. *Relatively small sample sizes may suffice if no overstated dollar units are encountered in the initial sample when the two-step approach is used.*

   The sample size required for the initial sample is dependent on the percentage of the total book record value represented by the minimum amount considered to be material. The higher this percentage, the lower the sample size required. Sample sizes of 50 to 100 may well serve the purpose. These sample sizes are usually much smaller than those required for the classical estimation sampling for variables plans.

2. *Unlike the traditional variables sampling method, the attributes approach provides an ability to establish an upper confidence limit even when no errors are included in the sample.*

   The method provides a highly probable maximum amount of overstatement in the population even when no overstated dollar units are found in the sample.

3. *Unlike some other forms of Dollar Unit sampling, the attributes approach provides two-sided interval estimates, with true confidence limits, for the total overstatement in the population.*

   As will be shown later, some of the other methods of evaluating Dollar Unit samples, provide upper bounds but not true confidence limits for overstatements.

4. *A major advantage of this method is that, while supplying exact confidence limits, it completely eliminates the problem of the nonnormal sampling distribution discussed in Chapter 3.*

   The form of the distribution of the values of the physical *sampling* units is not relevant since dollar units are sampled not physical units. The sampling distribution involved is the binomial distribution for which exact confidence limits can be established. Normal approximations are not necessary.

Combined Attributes–Variables Approaches

A number of other approaches to Dollar Unit sampling have been suggested, which use the same basic method of sampling dollar units, but which evaluate the sample directly on a dollar value basis. This is generally accomplished by assigning dollar values to (or "pricing") the selected dollar units.

Although not originally conceived of as a dollar unit approach, the Haskins and Sells[8] monetary sampling method developed by K. Stringer actually is a form of that technique. Another method, combined attributes and variables (CAV), which attempts the same type of approach, is discussed later.

The Haskins and Sells Audit Sampling Plan

The Haskins and Sells Audit Sampling Plan actually consists of two entirely different sampling plans.[9]

The sampling approaches are divided into two types. The first is called *numerical* sampling and will be recognized as an attributes (compliance test) sample. The second is called *monetary* sampling and constitutes an approach similar to, but *not the same as,* that described in the prior section, Dollar Unit sampling.

[8]This company currently operates under the name of Deloitte, Haskins and Sells.

[9]As described in the Haskins and Sells manual, *Audit Sampling, A Programmed Instruction Course,* 1970. The plan in the manual differs from the plans used in some prior years.

However, the Haskins and Sells manual notes that "most applications of the Haskins and Sells Audit Sampling Plan are based on the monetary amount of population items."[10]

Since the monetary sampling plan may not meet the requirements of the compliance test, the numerical method is supplied for that purpose.

Numerical Sampling

The numerical method outlined provides for the selection of *physical* sampling units by systematic sampling. The approach used is essentially the same as that outlined by the AICPA in the series *An Auditor's Approach to Statistical Sampling* using short-cut methods in lieu of tables.

Basically the plan provides for the selection of a sample size such that if *no* errors are found in the sample, the upper confidence limit will not exceed the MTER.

The sample is selected using systematic sampling techniques where the sampling interval is

$$i = \frac{N}{n}$$

Thus, if no errors are found in the sample, the compliance with the internal control system can be deemed to be acceptable, since there is a high probability that the actual error rate in the population does not exceed the MTER.

The method used for determination of sample size and for evaluation of sample results when errors are found in the sample is unique. The method used is based on a probability distribution referred to as the Poisson distribution, which provides approximations of binomial distribution probabilities in a simple fashion. A method for adjusting for the effect of a finite population is provided.

Certain reliability factors, or R values, are provided. These factors are equal to

$$R = np$$

where n = sample size
p = rate of occurrence[11]

[10]Ibid., frame 2-1, page 43. It is to be noted that more recently Haskins and Sells have added a series of traditional estimation sampling programs to their auditape system. These methods are described in M. S. Newman, *Financial Accounting Estimates through Statistical Sampling by Computer*, John Wiley & Sons, Inc., New York, 1976.

[11]The manual states that "in this course, the tolerable upper limit of the actual error proportion is designated by the symbol *P*. . . ." (Haskins and Sells), op. cit., frame 2-10, p. 52.

A few of these factors for different confidence levels are provided below:

Confidence Levels %	Reliability Factors (R)
80	1.6
86	2.0
90	2.3
95	3.0

These factors have been determined for the given probability (confidence level), where for a zero rate of occurrence in the sample, the upper confidence limit will be equal to the selected value of p. Thus, for instance, for the 95% confidence level, the reliability factor equals 3. Therefore,

$$R = np = 3$$

The sample size required to achieve an upper limit equal to an MTER of 5% is computed as follows

$$R = np = 3$$
$$p = .05 \text{ or } 5\%$$
$$n = \frac{R}{p}$$

Then, the sample size requirement is 60. This can be confirmed by reference to page 452 of the *Handbook of Sampling for Auditing and Accounting*, where for the table with 0% in the sample and a 95% confidence level, for a large population (100,000 and over) a sample of 60 would produce an upper limit of 4.9%.

If *any* errors are found in the sample, the upper limit obviously is now more than the MTER (in this case 5%). A new *upper* limit is then calculated by the Haskins and Sells sample evaluation process.

To understand the evaluation method, it is necessary to examine the upper limits when varying numbers of errors are found in the sample.

To illustrate this point, it will be assumed that the MTER is 3%. This would require a sample of 100 for the Haskins and Sells method, using a 95% confidence level.

An extensive set of tables of R and p_e factors is included in the Haskins and Sells manual. Factors are provided for confidence levels ranging from 50 to 99% and for numbers of errors from zero to 99. The upper portion of the table provides R and p_e for overstatement errors while a lower section provides factors for understatement errors.

For instance, the first few factors for overstatement for the 90% confidence level are given as:

Number of Errors in Sample	Precision Adjustment Factors (p_e)
0	2.3 (R)
1	1.59
2	1.44
3	1.36
4	1.32
5	1.26
.	.
.	.
.	.

To obtain a revised estimate of the upper precision limit, the precision adjustment factors (p_e) are accumulated up to that required for the number of errors found in the sample.

For instance, assume that a sample of 100 was used in order to have an MTER of 3% as the upper confidence limit at the 95% confidence level. Further, assume that the audit discloses three errors in this sample.

Since errors were found in the sample, the revised upper limit will be higher than the MTER (3% in this instance). A new upper limit must now be determined.

The Haskins and Sells manual gives the formula for the computation of the new upper precision level p' as follows:

$$p' = \frac{i(R + p_e)}{N}$$

However, since

$$i = \frac{N}{n}$$

The formula reduces to

$$p' = \frac{R + \Sigma p_e}{n}$$

Thus, from Table 7-2 of Haskins and Sells factors, the 95% confidence level precision factors for 1, 2 and 3 errors are 1.75, 1.56 and 1.46.

$$R = 3.0 \qquad \text{(95\% confidence level)}$$

Table 7-2 **Upper Confidence Limits for Sample of 100 for Large Population**
(95% One-Sided Confidence Level)

Number of Errors in Sample	Upper Confidence Limit (Percent)	Upper Confidence Limit: Increases from Previous Number of Errors	Haskins and Sells Precision* Adjustment Factor (p_e)
0	3.0	3.0%	3.0 (R)
1	4.7	1.7	1.75
2	6.2	1.5	1.56
3	7.6	1.4	1.46
4	9.0	1.4	1.40

*The precision adjustment factors represent values of np and must be divided by n to obtain percentage values. The symbol p_e is used to represent the precision adjustment factors.

Hence,

$$p' = \frac{3.0 + (1.75 + 1.56 + 1.46)}{100}$$

$$= \frac{7.77}{100} = 7.77\%$$

This can be confirmed by reference to the tables in the *Handbook of Sampling for Auditing and Accounting*, 2d ed., p. 379, where the upper limit is 7.6%. The disparity between these figures arises out of the fact that the Haskins and Sells factors are Poisson approximations and apply to infinite populations.

Of course, this approach based on one-sided upper confidence limits gives no possible ability to demonstrate that a population is unacceptable, since no lower limits are determined.

Once again, as detailed in the discussion of the AICPA compliance test method, if no errors are found in the sample, there is no question that the population is acceptable. However, if any errors are found, the result is indeterminant since *either* the lower limit is above the MTER and the population is unacceptable *or* the population is acceptable but due to an inadequate sample size the upper limit is above the MTER.

Monetary Sampling

For *monetary* sampling, the above numerical sampling techniques are combined with a sample selection technique similar to that described as Dollar Unit sampling.

As in Dollar Unit sampling, the Stringer method relates only to overstatements. Noting that understatement, for the reasons previously discussed under Dollar Unit sampling, would not necessarily be detected or fully reflected by their method, they suggest the use of tests of a reciprocal population to detect understatement.

The sample is selected in a manner similar to that previously described for Dollar Unit sampling. A systematic sample of *dollar units* is obtained.[12] The sampling interval is selected by using the formula:[13]

$$J = \frac{MP}{R}$$

where J = composite judgment factor (Dollar Unit sampling interval)
MP = upper precision limit
R = reliability factor for specified confidence level (from Table 7-2)

The upper precision limit (MP) is equated to the maximum amount not to be considered material. This, of course, is just below whatever would be considered a material amount.

The value P can now represent the proportion of the total dollar value which is construed to constitute a minimum material error; if M is interpreted to be the total value of the population, then

$$P = \frac{MP}{M}$$

For instance, assume a total value for the population of \$500,000 and a material amount of \$25,000; then:

$$P = \frac{\$25{,}000}{500{,}000} = .05, \text{ or } 5\%$$

Thus, more than 5% of the *dollar* units must be overstated to show the existence of a material error.

The selected *physical units* which include the chosen dollar units are audited and errors for these units determined.

[12]In numerical sampling, a systematic sample of physical sampling units was obtained. In monetary unit sampling, a systematic sample of dollar units is used. This provides a sample for which the probability of selection is proportional to the size of the physical sampling unit.

[13]No finite population correction factor is provided. However, since the population consists of dollar units which are numerous, in most cases this constitutes no problem. Since $R = NP$, then $J = MP/NP = M/N$. The symbol J replaces I as used in numerical sampling.

The new (revised) upper limit of overstatement is now computed from

$$MP' = JR + E$$

where MP' = upper (precision) limit based on sample results (revised upper precision limit)
J = sampling interval in dollar units (composite judgment factor)
R = reliability factors
E = sum of "adjusted" errors from sample

It will be noted that when no errors are found in the sample

$$MP' = JR$$

but $$J = \frac{MP}{R}$$

therefore $$MP' = MP$$

Thus, if no errors are found in the sample, the upper limit becomes an amount less than a material error and the account balance is acceptable.

However, if *any* errors of overstatement are encountered in any of the physical sampling units, the upper limit automatically is more than a material amount.

The Haskins and Sells manual at this point notes that if overstatement errors are found (without offsetting understatements) "the auditor at this point would have to decide whether the MP' was acceptable and if not, what further action to take (such as establishing a reserve)."[14]

In effect the auditor might reconsider his original determination of this amount of a material error (MP) and perhaps raise it, so that the new amount will be greater than the determined MP', according to the manual.

Any errors of overstatement on individual sampling units are multiplied by the ratio of the original value of that unit to the sampling interval (J) and in turn by the appropriate precision adjustment factor (p) (see Table 7-2 above) for that error. Since there is a decline in the value of p_e as the number of errors increase, to be conservative the errors are arranged in order, with the largest first.[15] These values are then added to the MP value. The value p is based on the *upper* limit for the percentage of such errors.

[14]Haskins and Sells, op. cit., frame 3–94, p. 120.

[15]There is no basis in statistical theory for this ordering of errors. Apparently it is done to provide a more conservative (higher) revised upper limit (boundary).

This approach results in a very conservative overstatement of the upper limits of the monetary errors.

Thus, the estimate provided for the revised upper confidence limit is *not*, as in estimation sampling, an estimate of the *probable* maximum total error, but a value which is *possibly* the maximum given certain very conservative assumptions. It is virtually certain to be *higher* than the upper limit achieved by ordinary sampling techniques.

For monetary sampling, the upper limit is no longer a confidence limit. It has been suggested that this upper limit be referred to as an upper bound or boundary.

However, the Haskins and Sells manual notes that "obviously, the monetary precision used in designing a sample is a relative rather than an absolute judgment, since there are no authoritative criteria for exact computations; therefore, further consideration of the upper limit after adjustment is justifiable."[16]

This seems to be a suggestion that if the sample data do not support the conclusion that the situation is acceptable then it would be appropriate to change the criteria.

The computation of the revised upper limit value is obtained by adding certain "adjusted" error amounts, based on errors in the sample, to the *MP* value.

The method of determining these "adjusted" error values involves weights for individual errors based on the upper limits for projected errors of overstatement, offset by *lower* limits for understatement errors.

The method of establishing a revised upper limit (boundary) involves the determination of the adjusted errors, which consist of all errors for values in the top strata plus certain weighted errors for all other strata.

It will be observed that if any physical sampling unit has a value equal to or in excess of the sampling interval (in dollar units), all such physical units will be included in the sample. Thus, all physical units with such high values will constitute a top stratum to be sampled on a 100% basis.

Thus, if the sampling interval is $10,000, all physical units with values equal to or greater than that amount will be included and will be audited. This is referred to in the Haskins and Sells literature as the top stratum.

Here is an illustration to explain the method in detail. In the audit of the XYZ Company, monetary sampling was used to test the reasonableness of an inventory balance. The monetary precision *(MP)* chosen for

[16]Haskins and Sells, op. cit., frame 3–94, p. 120.

this test was \$20,000. This is the maximum amount *not* considered to be material.[17] The total dollar value of the inventory (M) was \$1,000,000. This was the book record value. It was decided to use a 95% confidence level. The R factor of a 95% confidence level is 3 (see Table 7-2 or the Haskins and Sells manual).

These facts can be summarized as follows:

$$\begin{aligned} \text{population value} &= M = \$1{,}000{,}000 \\ \text{monetary precision} &= MP = \$20{,}000 \\ \text{confidence level} &= 95\% \\ \text{reliability factor} &= R = 3 \end{aligned}$$

The Dollar Unit sampling interval is

$$J = \frac{MP}{R} = \frac{\$20{,}000}{3} = \$6{,}667$$

Using the method described under the attributes approach to Dollar Unit sampling (page 114), the sample of dollar units and physical units was selected. Since the population of dollar units (M) was 1,000,000 and the sampling interval was 6665, a sample of about 150 can be expected. These physical sampling units were audited and four errors were found as shown below (the parentheses indicate an understatement error):

Stock Item Number	Book Record Value V	Error Amount e
1245	\$10,000	\$515
1593	5,000	200
1792	2,000	400
1999	3,000	(200)

The first step of the two-step procedure is to expand these errors to estimate the total population errors indicated by the sample. This is accomplished by multiplying each error amount by the ratio of the sampling interval to the value of the item in which the error occurred, for all errors except those in the top stratum

$$E_o = \frac{J}{V} e$$

where V is the book record value of the physical sampling unit.

[17]This value (MP) is the same as that previously referred to as the minimum amount considered to be material (MM).

For instance, item #1593 had a total value of $5,000. Since the sampling interval J was $6667, the sampling rate of stock items with a value of $5000 is $6667/$5000, or 1.33. This factor is assigned the symbol i_s. The expansion of the error in this stock item to the whole population is then $200 times 1.33, or

$$E_o = e(i_s) = (\$200)(1.33) = \$266$$

where E_o = projected population error for given error not in top stratum
e = error in selected sampling unit

$i_s = \frac{J}{V}$

V = book record value of sampling unit

For this sample the expansion of all sampling unit errors, except for top stratum errors, is shown below.

Stock Item Number	Book Record Value V	Error Amount e	Ratio i_s	Estimated Population Errors E
1593	$5000	$200	1.33	$ 266
1793	2000	400	3.33	1332
1999	3000	(200)	2.22	(444)

There is no need for such an expansion for errors in the top stratum of sampling units since all will be included in the sample. The population total for items in this stratum will merely be the total of those in the sample. The projected error for the population will then be the projected errors from the lower values plus those for the top stratum.

$$E = \Sigma E_o + \Sigma E_t = \$1598 + \$515 = \$2113$$

From this total may be subtracted any adjustments made as a result of the errors found in the sample.

In a second step, these figures may be used to obtain a *revised upper limit* (p'), by multiplying these projected population errors by the precision adjustment factors (p_e) from Table 7-2 (or the table in the Haskins and Sells manual).

In this example there were two overstatement errors, in other than the top stratum, and one understatement error. When there is more than one error or overstatement (or understatement), the errors are ranked and ordered according to magnitude. The ranking and ordering of the two understatement errors in this example are shown below:

Stock Item Number	Error (Ordered Sequence)	Rank
1792	$400	1
1593	200	2

The precision adjustment factors (p_e) from Table 7-2 (95% confidence level) are 1.75 for the first overstatement error and 1.58 for the second overstatement error. The estimated population errors (E_o) from the previous calculation are then multiplied by these factors in their ordered sequence. A separate table of precision adjustment factors (i_s) are supplied in the Haskins and Sells manual. For one understatement error at the 95% confidence level the factor is .05, as shown in Table 7-3.

To the net total of $2723 must be added the errors in top stratum sampling units:

$$E' = \Sigma E_o + \Sigma E_t = \$2723 + \$515.00 = \$3238$$

This value now represents the increase in the previously selected *MP* (based on no errors in the sample). The revised upper limit (bound) will be equal to

$$MP' = E' + MP = \$3428 + \$20{,}000 = \$23{,}238$$

It will be noted that if *any* overstatement errors are found, the revised upper limit will be in excess of the minimum amount considered to be material (in this instance $20,000). The first increment of the upper limit of errors ($20,000), which is always added to the projection

Table 7-3

Stock Item Number	Estimated Population Errors E_o	Precision Adjustment Factors p_e	Precision Adjusted Errors E'
	Overstatement Errors		
1792	$1332.00	1.75	$2331
1593	266.00	1.56	415
	Understatement Errors		
1999	$ 444.00	.05	(22)
		Net Total	$2723

of the actual errors found in the sample, in effect makes the assumption that the physical sampling units composing this increment were 100% overstated.

It must be emphasized that the fact that the upper limit (bound) is higher than the minimum amount considered to be material, as will happen when any errors of overstatement are found, does *not* mean that there is a material error in the balance tested. It means that it is possible but no probability can be attached to that possibility. This is particularly true since the upper limit produced will always be higher than the true confidence limit when errors of overstatement are found. Without a method for fixing a lower limit it is not possible to prove that a material error exists.

An additional procedure provides the *frequency* of error (attributes sampling) from the Dollar Unit sample when both numerical and monetary evaluations are required. The procedure is outlined in detail in the Haskins and Sells manual.[17a]

Roberts expresses the Stringer method of establishing the upper bound of population errors by a formula which, when translated into the symbols used below, provides a better understanding of the process.[18]

$$MP' = M\left(P + p_1 \frac{E_{o(1)}}{V_1} + p_2 \frac{E_{o(1)}}{V_2} + \cdots + p_i \frac{E_{o(i)}}{V}\right) + E_t$$

where $p_1, p_2, \ldots, p_i$ are the precision adjustment factors (p_e) *in order of rank.*

Summary of Limitations

This approach is similar to that used by the AICPA for compliance tests in that only upper limits can be established. Since a lower limit is required to establish that in fact the errors are material, only accepta-

[17a]Haskins and Sells, op. cit., pp. 122ff.

[18]D. M. Roberts, *Statistical Auditing,* American Institute of Certified Public Accountants, New York, 1978, p. 121. The original version of the formula as stated using Roberts' symbols is (for overstatement errors only):

$$E_o = Y\left\{P_u(0) + [P_u(1) - P_u(0)]\frac{d_1}{y_1} + [P_N(2) - P_u(1)]\frac{d_2}{y_2} + \cdots + [P_u(k) - P_u(k-1)]\frac{d_k}{y_k}\right\}$$

where

$P_u(1), P_u(2), \ldots, P_u(k)$ are the upper limits of error rate for $1, 2, \ldots, k$ errors in the sample.

P (0) is the upper limit rate of occurrence when no errors are found in the sample.

$d_1, d_2, \ldots, d_k$ are the dollar values of errors found in the sample in order of rank (largest to smallest).

$y_1, y_2, \ldots, y_k$ are the book record values of the physical sampling units in which the above errors were found, also in order of rank.

And, since $MP' = \Sigma E_o + \Sigma E_t + MP$, the above version of the formula results.

bility can be firmly established. Since the estimates of upper bounds are very conservative, there may be numerous false alarms.

The upper limit method used here merely establishes whether it is *possible* (not probable) that a material error could exist in the account balance. Failure to establish that it could not exist results in an indeterminate situation, since it is not known whether the material error actually exists. A further limitation is that the method admittedly applies to overstatement only.

As previously observed, if any errors of overstatement are found, the first increment toward the revised (and higher) upper limit or bound is the value *MP* which is added to the projection of the actual errors found and which results in the assumption that the physical sampling units composing this increment were 100% overstated.

This, however, involves the assumption that a physical sampling unit can never be overstated by more than its full value and there are many situations where this can happen. If credit balances can exist for given data, the statement of a value as a debit amount when it should have been a credit balance would provide an error of more than the stated balance.

Finally, the method will not include in the sample any accounts amounting to zero or with credit balances although these may be in error.

In general, the method provides an extremely conservative upper bound as compared with the upper confidence limit computed by other methods. If no errors are found, this conservatism does no harm. The population has a high probability of not including a material error, depending on the confidence level used.

The difficulty arises when errors are found. With excessively high upper bounds and with no ability to establish actual materiality of errors without lower limits, unnecessary concern and work may result.

Of course, all of the previously mentioned limitations of Dollar Unit sampling apply to this method as well.

Other Dollar Unit Methods

The Haskins and Sells method may be seen to be a combined attributes and variables approach (CAV). Several such methods have been proposed.[19]

[19]J. L. Goodfellow, J. K. Loebbecke, and J. Neter, "Some Perspectives on CAV Sampling Plans," parts I and II, *CA Magazine*, October and November 1974. R. Anderson and L. Teitelbaum, "Dollar Unit Sampling," *CA Magazine*, April 1973, pp. 30–39. W. F. Wilkerson, *HEW CAS and Dollar Unit Sampling*, footnote, HEW Audit Agency, U.S. Dept. of Health, Education and Welfare, June 1975. D. Leslie, L. Teitelbaum, and R. Anderson, *Dollar Unit Sampling*, Pitman, New York 1979.

All CAV methods involve some method of assigning a dollar values to or "pricing" the increments in the upper bound arising from finding one or more errors. Commonly these methods assume that the basic upper limit resulting from finding no errors in the sample is assigned an overstatement value of one dollar to each dollar unit for the percentage of the population indicated by the upper limit. The increase in the upper limit will then be assigned dollar values derived from errors found in accordance with the errors in the physical units for which errors are detected.

It is to be emphasized that the upper limits produced by any of these methods when errors are found are *not* upper confidence limits but upper bounds which will overstate it by a possibly considerable amount overstatement. It is not an estimate of the actual maximum population overstatement but rather how much it might be at the most based on conservative assumptions about the errors. These methods produce no lower confidence limits or bounds. The methods, thus, cannot be defended using classical statistical theory, as an interval estimate of the actual population value of the overstatement.

These methods also provide means of providing for understatement as an offset to overstatement. Generally the offsets are based on the *lower* limit estimate of understatement in order to be most conservative.

Generally speaking, since the upper bound of the overstatement when errors are found is so conservative that if the upper bound is more than the dollar amount considered to be material, nevertheless there is a high but unspecified probability that there is no material overstatement.

However, all of these methods collapse when errors are found and the upper bound is *more* than the selected material amount since *no* realistic estimate can be established for what the maximum overstatement is *nor* whether an overstatement actually exists, since no lower confidence limit can be established.

Chapter 8

Problems of Sample Selection

For sample results to be meaningfully evaluated on a statistical basis, it is essential that the units composing the sample be obtained from the population using probability selection methods. Since the principle of evaluation of sample projections is based on the concept of a predictable sampling distribution, the method of selection of the sampling units must be in accordance with the principles of probability sampling.

The sampler may be well intentioned, the sample wisely and carefully selected, yet the sample created by judgment or selective methods may defy the possibility of an objective and mathematical evaluation.

Questions of cost, convenience, or limitations of time *cannot* be allowed to interfere with the use of the correct method. Even though the technique, at times, may be considered awkward or impractical, unless the appropriate method is used *without compromise*, the sample projection cannot be evaluated statistically.

This does not imply that samples selected for audit by an auditor on a judgment basis are of necessity incorrect. In fact, it is conceivable that such a selective judgment sample may give results which better represent the facts than a probability sample. However, there is no way of knowing how far wrong the judgment sampler may be in the sample evaluation or to recognize the accuracy of the determination. On the other hand, the sample created by probability sampling methods provides an estimate (in probability terms) as to how far the sample results might be from the results of a 100% audit.

It is important to emphasize that the required random selection for a statistical (probability) sample does not in any way restrict the auditor's ability to audit records other than those included in the sample.

If an auditor, on the basis of knowledge, experience, suspicion, or intuition, feels that certain transactions or entries require investigation, then by all means the auditor should investigate.

However, any item examined for any reason other than its inclusion in the randomly selected sample cannot be included in the sample projection. It is excluded from the sampled population and, if appropriate, may be added to the sample *projection*.

THE UNRESTRICTED RANDOM SAMPLE

The most widely used sample selection method,[1] unrestricted random sampling, requires that all units in the population have an equal probability of inclusion in the sample.[2] All possible samples (combinations of sampling units) in the population must also have an equal chance to be selected.

To achieve an unrestricted random sample only a few methods are available. These methods include the thorough mixing of the records, a technique not practical in the auditing situation, random number sampling, and systematic sampling.

The population to be sampled in the audit situation is quite different from those sampled in other areas. The accounting population is always finite and usually consists of records concentrated at a limited number of locations, most usually at one location. It is also to be noted that the population to be sampled in the audit situation is most often the population of audited values or of errors, an unknown population, since a 100% audit would be necessary to disclose that population.

PROBLEMS OF RANDOM NUMBER SELECTION

The selection of sampling units using random number selection methods assumes a source of random numbers and a series of documents numbered, or construed to be numbered, in such a fashion that there is a one-to-one correspondence between the selected random number and the selected sampling unit. Thus the specification of a random number must call for the selection of a single sampling unit and only that sampling unit.

The necessary random numbers can be obtained from any of a number of published tables of random digits.[3] Alternatively, a suitable ran-

[1]Other sampling methods may provide other criteria for sample selection, such as selection with probability according to size or merely a known probability of selection.

[2]For finite populations, this becomes an equal probability that the remaining sampling units be included in the sample after each sampling unit is drawn.

[3]For instance, *One Million Random Decimal Digits*, Rand Corporation, Free Press, New York, 1955.

dom number generator programmed with a batch processing or time-shared computer system can more efficiently supply lists of random numbers.[4]

The use of a table of random digits to provide the required random numbers requires a random start in that table and the selection of numbers beyond that point in a consistent fashion.[5]

The use of a random start in a random number[6] table meets the requirement that there be an equal probability that any given set of random numbers in the entire table be selected, thus providing an equal probability that any given sample will be drawn. Further, the use of a random starting point provides the auditor with a necessary element of surprise in the selection of records for a test so that the items to be audited cannot be foretold. This is an important protective feature provided by random number sampling.

Groups of random digits of sufficient magnitude (number of digits) to cover the range of the numbers of the documents are obtained. The table of a random digits may be read in any direction, but it is suggested that a reasonable and consistent pattern of selection be followed rather than a haphazard or capricious procedure. The use of a random start will supply objectivity to the choice of sampling units for inclusion in the sample and prevent any bias on the part of the auditor, *conscious or subconscious,* from affecting the selection.

When random numbers are generated using computer terminals, the random start is usually provided by an initiator number supplied to the computer by the auditor. This initiator number must be chosen randomly and usually within certain restrictions provided by the computer program. However, some time-sharing terminal programs provide their own initiator number by resort to a "clock" within the computer.

It is important to note that the use of random number selection methods has the same effect as giving the sampling units a thorough, physical, mixing. This eliminates any effect that the original sequence or arrangement of the records in the file may have had.

Since there is great diversity in record-keeping systems, it is inevitable that problems will arise in trying to sample each system using random number selection.

[4]See *Random Number Generation and Testing,* International Business Machines Corp.

[5]For a detailed description as to how a table of random digits may be used see H. Arkin, *Handbook of Sampling for Auditing and Accounting,* 2d ed., McGraw-Hill Book Company, New York, 1974, chap. 3.

[6]Although usually referred to as random number tables, these tables, in most cases, are compilations of random digits which may be combined to create random numbers.

Unusual nonsequential numbering systems, or filing systems where a document cannot be located by number, may make the use of the numbers on the documents impossible or impractical. Or the documents may not be numbered at all. In such cases, it may be desirable to imply numbers for the documents rather than to attempt to physically number them.

This implication of numbers for unnumbered documents may be achieved in a variety of ways. It may be possible to imply numbers by counting records where the count specifies the record number. For manual record-keeping systems this approach is practical only for small populations. For computerized data, on tapes or cards, record numbers can be created by computer counting. With this method, when the count reaches the selected random number, the specified record is produced by the computer.

PAGE AND LINE NUMBER SELECTION METHODS

For manual systems, where listings or trial balances of some sort are available, selection may be accomplished by the page and line number selection method. Where this technique is applied, a random number of from one to the total number of pages is selected and paired with a line number of from one (to the *maximum* number of lines possible or number of lines used) per page. That page number is located and the line number is found by counting. This specifies the sampling unit to be included.

If the paired page and line numbers selected indicate an unoccupied line, *both* page and line numbers must be deleted and replaced by a new pair. It is not proper to select a new line number only, since this would result in an equal probability of selection for each page rather than for each line item, as required.

If the sampling unit occupies more than one line on a page, in counting to reach the "line" specified, the count would be of groups of lines or sampling units. For instance, if the test is of a payroll to determine errors in payments to individuals, where individuals may occupy varying numbers of lines (one for regular pay, one for overtime pay, etc.), the count would be of employees, not lines.

If listings are not available, if may be possible to simulate page and line number sampling. For instance, if the records are in numerous file cabinet drawers, a random number can be drawn to represent the file cabinet drawer and another number to specify the record within the drawer. For instance, if number 27 is drawn for the drawer and 32 for the documents within the drawer, the 27th drawer would be selected on the basis of some previously established counting system and the documents within counted until the 32d was found.

If there are too many documents within a drawer to make counting such records feasible, arbitrarily placed numbered file dividers can be used. These should be spaced at approximately equal intervals. A random number may then be drawn for the file divider number and a second number for the document within that group. The random number for the document should be between one and the maximum number of documents behind my file divider. Such a maximum can be assured by spacing the dividers to make one group larger than any other. The number of documents behind that divider will be determined, and if a random number pair is selected beyond the last number in the specified group, another pair of numbers will then be drawn.

The principle of page and line number sampling can be carried a step further if necessary. For instance, assume a business organization which has offices at a number of locations with a considerable number of employees at each location. The auditor wishes to select a sample of the line items on the weekly payrolls for audit.

The selection process may proceed by assigning numbers to the say, 20 offices, and numbers to the weekly payrolls (from 1 to 52) at each location. The sample may then be selected by drawing a random number from 1 to 20 to designate the office, and one from 1 to 52 to designate the week. A number of from 1 to the total number of pages on the largest payroll will designate the page, and one from 1 to the maximum possible numbers of lines on any page will designate the line number.

A set of numbers which might be drawn is shown below:[7]

Sampling Unit	Office	Week	Page	Line
A	18	03	9	22
B	01	36	5	49
C	11	1	1	1

Thus, for sampling unit A payroll 3 at office number 18 would be selected, page 9 chosen, and the item on line 22 on that page becomes the first sampling unit.

For sampling unit B, when the 36th payroll at office number 1 is examined and page 5 is found, it is possible that there may be no 49th line on that page for some valid reason. In that case, all of the numbers

[7]The assumption is made that there are not more than nine pages on any payroll and not more than 50 lines on the page. If there are more than nine pages in *any* payroll for this organization a two-digit page number would be required.

selected for that sampling unit, office, week, and page included will be discarded and a new set selected.

It is important to note that selection by this method is a true unrestricted random sample of all line items on all payrolls at all locations. This is *not* the same as drawing a separate sample (proportionate or otherwise) from each payroll independently and combining the results. The latter selection process produces a stratified sample and must be treated as such.

SPATIAL SAMPLING UNITS

In some sampling situations, there may be a widely distributed heterogeneous group of items to be sampled. In one case, it was desired to estimate the total value of goods in process in a factory. The material to be evaluated ranged from raw materials to completed items. These materials were widely intermingled on large factory floors.

The most feasible sample selection technique in such a situation would be to use a spatial sampling unit. The factory floors were theoretically divided into small, uniform areas. These spatial sampling units were theoretically numbered. A drawing of random numbers designated the spatial sampling units to be included in the sample. Each sampling unit was located and evaluated by counting, pricing, and extending the values of the items contained within the sampling unit (spatial unit) to establish a value for that unit. The sample size in this case was the number of spatial sampling units included.

In another situation, where it was necessary to establish intercompany payments of rentals for the joint use of poles by the telephone company and the local utility, the area involved was divided into segments of 1050 feet by 1650 feet, using public utility maps. A random number selection determined which of these areas were to be included in the sample. Each selected area was completely evaluated.

When spatial sampling units are used, it is not unusual to encounter "empty" sampling units. Thus, the selection of the spatial areas on the factory floor may result in an area which contains no inventory items. The selection of the areas on the utility map may result in an area which has no poles. Often it cannot be determined that a sampling unit is empty until it is examined. Ordinarily, these empty units are not discarded but their value is recorded as zero for a variables projection. There may well be other such empty units in the population. If the number of empty units is known in advance, any empty unit selected may be discarded and another spatial sampling unit drawn in its place. But if this is done the number of empty sampling units must be subtracted from the total population. This last method (replacing empty units) is preferable for a variables estimate, since it is likely to result in a lower standard deviation for the population and thus a smaller sampling error.

Other unusual sample selection problems may be encountered. In one instance, an audit of deliveries made to a large number of locations was undertaken. There were daily deliveries at a specified time to each site. It was desired to project the dollar value of improper or missing shipments. The sampling unit in this situation was designated as a site-day.

A random number selection of site-days was made with one number representing a site and another a day during the audit period. The auditor was at the selected site on the selected day to check on the delivery and establish the value of any error in the shipment. It is to be noted that the population size (N) was the total number of site-days (sites times days) for the audit period.

DUMMY SAMPLING UNITS

In the previous sections it was noted that when spatial sampling units are used, there may be instances in which some of the selected sampling units are empty. Under certain circumstances such empty or dummy sampling units may be included in the sample deliberately. Thus, when the exact population of sampling units is not known, but it is known to be no more than a certain figure, the inclusion of the empty units provides a population of a known size. The size of the population must be known to project a total dollar value. Of course such empty or dummy units will each have a dollar value of zero.

When dealing with documents, for example, each of them may contain a different number of line items and the total population of the line items (sampling units) may not be known. The sampling units may be selected using the page and line number approach. However, for variables (dollar value) sampling, if it is desired, a population of a known size may be created if when the line number drawn for a particular page is found to be unoccupied, it is given a value of zero.

For instance, if 2000 invoices are to be sampled but the sampling unit is the line item on the invoice, random numbers from 1 to 2000 can be drawn for invoice (page) selection. If it is established that no invoice has more than 50 line items, random numbers between 1 and 50 may be drawn for the selection of the line item on the selected invoices. Thus, if invoice number 1123 is selected and line 22 is the designated line, it may be found that there is no occupied line 22. There may be only 15 line items on that invoice. In that case, the value for that item in the sample becomes zero regardless of whether the true (audited) value or the amount of the error is to be estimated. The population size then becomes the number of invoices (in this case 2000) multiplied by the maximum number used for the line item selection (in this case 50). In this example the sampled population is now known to be exactly 100,000.

SYSTEMATIC SAMPLING

Systematic sample selection involves the selection of sampling units at intervals beginning with a randomly selected unit. In the most commonly used method this is a fixed interval. The sampling intervals may be obtained by dividing the population size by the sample size. This results in the sampling interval (k). The starting point (first sampling unit to be included in the sample) is obtained by drawing a random number between 1 and k.

This type of sample selection technique has long been used by auditors as a simple and objective method of sample selection for audit test purposes. However, the method has usually been applied without resort to a random start.

It is to be noted that while *statistically* unbiased, a sample drawn in this manner may be biased from the audit viewpoint. To be considered unbiased in the statistical sense, the only requirement is that the point estimates (in auditing usually proportions or averages) obtained from all possible samples of a specified population average to the value obtainable from a complete coverage (100% sample) of all sampling units in the population.

However, if a pattern exists in the ordering of the sampling units upon which the sample selection is based, a systematic sample may be produced which is biased from the audit viewpoint. Thus, if the auditor is sampling line items on a payroll when that payroll has a fixed pattern, such an audit bias may be introduced. For instance, if every 10 persons on the payroll are supervisors, a systematic sample with a sampling interval of 10, or any multiple of 10, will produce a sample containing only supervisors or containing no supervisors.

While such a sample is unbiased in a statistical sense, in that drawing all possible samples by such a method would produce point estimates which would average to the parameter of the population; from the audit viewpoint such a sample would be biased (noncross-sectional) and useless.

In this case an alert auditor might discover and forestall the effect of the payroll pattern by selecting an interval which is not a multiple of 10. In many instances, however, the pattern in the population is not obvious and would not be apparent from the sample results, especially if the pattern is not perfect. This inability to discover an existing pattern in the population is particularly insidious in auditing, since the population values are not known and cannot be determined without a 100% detailed audit. Since the audited population cannot be scanned for such patterns, and the existence of such patterns cannot usually be learned from the sample, a bias of an audit nature may result.

When documents or records are being sampled for an audit examination to provide an estimation sample to project dollar values, any pattern (even though not perfect) in these values (usually audited values) may cause an underproportion or overproportion of high- or low-value items to be included in the audit.

Patterns in document files are not unusual. For instance, if the file is numbered, numerical spaces may have been left to include new documents at appropriate points. In alphabetically arranged files, spaces may have been left to provide for new documents in their proper alphabetical sequence.

The effect of such patterns would be less serious if the statement of the sampling error allowed for these variations.

However, when a single systematic sample with a random start and a fixed interval is used, it may not be possible to compute the sampling error of an average or total projected from such a sample due to an inability to obtain a valid estimate of the population standard deviation from such a sample.

It is known that under certain conditions the sampling error of an average or total known systematic sample may be smaller than, the same as, or larger than a random number (simple random sample).[8]

If the sampling units in the population are in random order, the sampling error of an estimate from a fixed-interval systematic sample will be the same as that from an equivalent random number sample. If the values being sampled are in ordered sequence (i.e., smallest to largest or vice versa) the sampling error for a systematic sample will be less than that for a random number selection sample. If there is a periodic pattern to the value of the sampling units, the sampling error of an average from a systematic sample will be larger than that obtained from a sample selected on a random number basis.

If the auditor has reasonable assurance that the population has no pattern, he or she can assume that the sampling error computed in the usual manner for an unrestricted random sample is close to that of the systematic sample. However, the justification for such an assumption is based on a judgment evaluation by the auditor and is not defensible as

[8]The sampling error of an average computed from an unrestricted random sample times the factor $1+(n\text{-}1)r$ will be equal to the sampling error of a systematic sample where r is the coefficient of correlation between successive observations in such samples.

objective. Further, it is to be noted that it is the audited values which are sampled and often the errors in the values rather than the values themselves. Since neither the audited values nor the errors in the population are available for examination, any assumption about the distribution of values being sampled is suspect.

However, as noted above, since in this field it is the unknown audited population values which must be considered, there is considerable risk in such an assumption.

For this reason, it is urged that if the audit projection is to be subject to challenge, as in a court of law, it is well to avoid the use of systematic sampling due to this possible criticism.

The only saving in effort, in the usual audit situation when sampling records, is that no list of random numbers must be prepared. After all, it is still necessary to locate the specified sampling unit by number or by counting.

Other difficulties are encountered when systematic sampling is used. If the population size is not known exactly, the sample size attained will be larger or smaller than anticipated. If the sample turns out to be too large, the auditor cannot terminate the systematic sample early since his sample would then be drawn only from the first part of the list or file. Since the items in the end of the file would not be included in the sample, the projection from the sample would not apply to that part of the file.

In such an instance, the entire sample must be drawn and then reduced by selecting a systematic or random number sample of the items in the systematic sample for exclusion from that sample.

Thus, in general, an auditor cannot terminate the audit sample at an earlier stage than originally anticipated for a systematic sample if a sample projection is to be made.

If the sample turns out to be too small because of an erroneous estimate of the population size, it may be enlarged by drawing another systematic sample with a new random start and an interval determined by dividing the population size (N) by the additional number of sampling units required.

If, in spite of these considerations, the auditor still wants to use systematic sampling, the sampling error can be calculated by using several selections at fixed intervals in creating the sample. With the use of several random starts, the sampling units may be drawn at an interval of $(N/n)r$, when r equals the number of subsamples to be drawn.

It is suggested that not less than 10 such subsamples be obtained and if separate records are maintained for each subsample, the sampling error may be determined by resort to replicated sampling.[9]

[9]See H. Arkin, op. cit., chap. 14.

Another method of overcoming some of the difficulties encountered in using systematic samples is the use of a randomly varied sampling interval, where the interval is created for each selection by use of a random number between 1 and $2k$.

This method will overcome the audit bias which might exist, and the sampling error computed from this type of sample, while not exactly equivalent to that of a random number selected sample, should be quite close to the true sampling error, if not precisely the same.

EXTRANEOUS SAMPLING UNITS

The population (field) is defined by the auditor as consisting of all sampling units about which the auditor *wishes* to draw a conclusion or make a projection. There is no requirement to include every record (sampling unit) in the defined population. The field may be restricted to contain only those sampling units in which there is an interest. Of course, if such a restriction takes place, the projection of the sample result applies only to those sampling units in the defined field.

If, for good audit reasons, the auditor decides that only a certain group of vouchers are of concern, or that because of considerations of materiality only those vouchers of over $100 should be sampled, the auditor is certainly privileged to do so. However, any conclusions will apply only to these limited groups of records. Thus, conclusions in the form of a projection from the sampling of those vouchers with values in excess of $100 do not apply to vouchers of lesser amount. Nevertheless, if there is a valid audit reason for confining the test projection, there is nothing wrong with doing so.

However, during the sample selection process it may happen that a document selected by the sampling technique will not be one of those included in the defined field.

In the event that an extraneous document is drawn by the sampling process, one not applicable to the field, it may be disregarded and another selection made to replace it by the same process. It must be noted that the population size becomes the total number of documents in the field less the number of such extraneous documents in the population.

If systematic sampling is used, the sampling interval should be applied by counting only those documents validly in the defined field.

MISSING SAMPLING UNITS

While technically a *frame* consists of a listing of all sampling units in the population, usage has broadened the definition so that the term applies to all the sampling units available to be sampled.

It is not at all unusual for sampling units not to be in the file when sought by the auditor. It is possible that they are being processed or have been temporarily withdrawn for other purposes. It is also possible that they have been withdrawn to keep the auditor from examining these documents. They may have been lost or destroyed.

Since the projection of a sample result applies to the frame and to the population only if the frame and population are identical, it becomes the responsibility of the auditor to establish that the frame and population are the same. In other words, the auditor must be sure that the population is intact. It is one of the advantages of random number sampling, which is applied only when the sampled documents are actually numbered, that this sampling technique may disclose the fact that sampling units are missing. When the auditor seeks the document with a number corresponding to that of the selected random number, it may not be there.

If the auditor cannot find a sampling unit with the required number, it is *not* proper to replace it by drawing another random number. The auditor should exert every effort to locate the missing document or establish a reason for its absence.

If the document cannot be found, and no adequate explanation can be established for its disappearance, a projection can be made from the sample of the proportion of the population of records which similarly would be missing for reasons unknown. This is a very important audit finding. If the missing document is important or is essential to establish the validity of a transaction, or if more than an insignificant number are missing, serious consideration must be given to the impact of such a conclusion on the ability of the independent public accountant to express an opinion on the fairness of the financial statements. The auditor will not be able to establish that compliance with the internal control system is adequate. In internal or governmental auditing such circumstances indicate a highly suspicious situation where, at a minimum, steps must be taken to prevent the recurrence of such events.

In order to accomplish a valid projection, it is necessary to establish an audit determination for *every* sampling unit included in the sample. If the information about a sampling unit is not available from the usual source, it is necessary either to obtain the information from alternative sources *or* to make an assumption about the validity of the sampling unit.

Thus, if a receiving ticket cannot be found for an invoice payable included in the sample, other assurance must be obtained that the invoiced material was indeed received. This alternative information must be as incontrovertible as the receiving ticket itself.

If a dollar value estimate is to be made, and such assurance cannot be obtained from any source, the sampling unit must be considered to have an error equivalent to its entire value. In attributes sampling (the compliance test) this situation would indicate the occurrence of an error.

VOIDED DOCUMENTS

Occasionally, the sample selection process may result in the drawing of a voided document. It is important to note that the auditor should not immediately select another sampling unit to replace it. An investigation should be undertaken to establish that the document has been *properly and validly* voided. A clerical error may have resulted in the voiding. However, it may have been retained in the file, when serially numbered, to account for the entire series. If this condition exists, another sampling unit may be selected to replace it.

If there is evidence that there is *not* a valid reason for the voiding of the document, it should *not* be replaced by another selection, but a projection should be made from the sample result as to the extent of such invalid voidings in the field.

ERRORS IN SAMPLING UNIT SELECTION

When probability sample selection techniques are used, if appropriate care is not exercised, sampling units may be included in the sample which are not actually designated for such inclusion, in the place of other properly designated units.

While obviously such errors should be avoided, if they are inadvertent they will not invalidate the sample. Such errors are merely random deviations on a random sample and thus produce a different but valid probability sample.

Nevertheless, errors of this type result in the possibility of a challenge, if there is a conflict over the validity of a sample projection, especially in a court of law. It may be contended that the apparent errors were deliberate and constitute an attempt to bias the sample by selecting the sampling units most favorable to one's objective. Since there is no way of proving that such errors were indeed inadvertent, such a challenge may be sustained.

For this reason it is essential that sufficient care be exercised in preparing the lists of random numbers and in selecting the sampling units corresponding to those numbers to avoid such failures. It is usually desirable to double check the selection process.

Chapter 9

Sample Data as Legal Evidence

Accounting and economic data based on sampling have been placed in evidence before courts of law and administrative commissions for many years.[1] A search of law literature relating to the admissibility in court of sample data arising from audit tests has disclosed no instance in which such a test by an independent public accountant has been contested in court.[2]

Recently, however, a projection by a governmental auditor based on a statistical audit sample was contested in a court of law.[3] The case related to a projection of overpaid medical claims from a sample and resulted in $3.5 million in federal payments being withheld from the state of Georgia.

The state of Georgia appealed to the courts for relief and presented several arguments, one of which objected to the use of a projection from a statistical sample rather than a claim-by-claim examination to determine the amount of the overpayment.

[1]Numerous cases relating to the violation of the Pure Food and Drug Act have involved the use of sample evidence. While in many cases sample accounting or other record data have been placed in evidence and accepted as admissible, the most widely known case is *Illinois Bell Telephone v. Illinois Commerce Commission* 414 Ill. N.E.2d 329 (1953). Supreme Court Record on Appeal, II. 714–718, 809–875.

[2]However, it is to be noted that this does not signify that there were no cases in which sample data was used since unless the sample data was contested there would be no mention of it in the literature as it would not be discussed by the court.

The only case which could be found relating to an audit test was *Ultramares v. Touche*, 225 N.Y. 170, 191–192, 174 N.E. 441, 449 (1931). In this case, the court stated that the audit test method was not in question since the failure to detect fictitious invoices by a test was irrelevant because "the clerical entry in the ledger of assets, plainly interpolated and suspicious on its face was excluded" from the examination.

[3]*Georgia Department of Human Resources v. Califano*, 446 F. Supp. 404 (1977).

The ruling by the judge stated:

> Georgia asserts that the Administrator's decision was arbitrary and capricious because the amount of overpayment was determined by use of a statistical sample rather than by individual claim-by-claim review. The court concludes that the use of statistical samples was not improper. Projection of the nature of a large population through review of a relatively small number of its components has been recognized as a valid audit technique and approved by federal courts in cases arising under Title IV of the Social Security Act.

Most cases involving the use of sample data as legal evidence have involved the use of public opinion or similar survey data. The admissibility of this type of sample data has been contested as being hearsay evidence. For this reason, such sample surveys have often been rejected by the courts.

However, it is generally accepted by the courts that business entries or records made in the ordinary course of business constitute an exception to the hearsay rule. This exception, it must be noted, merely makes the universe data and the sample data admissible and in itself does not meet all objections to the admission of sample data as evidence.[4]

The Federal Judicial Council's manual for judges states that while "the principal objection to the admission of samples and polls has been that such evidence is hearsay," nevertheless, "courts now admit samples and polls over the hearsay objection on the grounds that surveys are not hearsay or on the grounds that surveys are within a recognized exception to the hearsay rule."[5]

When the type of data sampled relates to factual rather than opinion matter, the courts have tended to look more favorably upon such sample data.[6]

To an increasing degree, audit sample data have been used to attempt to reclaim sums of money improperly collected by the organization audited. This type of situation has arisen with respect to claims for disallowances upon audit of a government contractor, a government agency (such as a state or county) which obtained or claimed funds from a superior government agency in a manner not in accordance with the

[4]Allan H. McCoid, "The Admissibility of Sample Data into a Court of Law: Some Further Thoughts," *UCLA Law Review*, 1957, pp. 233–250.

[5]*Manual for Complex and Multidistrict Litigation*, Federal Judicial Center, 1973, sec. 2.712.

[6]"First courts are more likely to admit sample evidence as to objective facts when they are sample evidence as to group or public opinion." F. Kecker, "Admissibility of Economic Data Based on Samples," *Journal of Business*, April 1955, pp. 114–127.

regulations for such payments, or in connection with attempts to reclaim tax payments where some of these taxes were not properly paid.

The massive quantities of documentation involved in modern litigation may well have led to the acceptability of sample data. The Federal Judicial Council's manual for judges notes that "the proper use of samples and polls as a means of facilitating proof in protracted litigation is receiving increased attention and has great potentiality." Further, the manual states that "scientifically designed samples and polls, meeting the test of necessity and trustworthiness, are useful adjuncts to conventional methods of proof and may contribute materially to shortening the trial of complex cases."[7]

The manual states, however, that "the burden of proof rests upon the offeror to show that the sample was selected in accordance with accepted principles of sampling so that it properly represents the universe."

The manual further notes that "such surveys usually are based upon 'probability sampling,' a system under which every unit or person in the universe has a known chance of being included. In this type of survey the results, by application of statistical principles, may be projected to the universe with a known 'sampling error.' This kind of error is to be distinguished from 'error in sampling' which is caused by improper, i.e. biased, selection of the sample."[8]

Thus, the legal emphasis is becoming focused on the use of properly designed samples. In a 1957 case the judge in admitting a survey emphasized that "the sample design utilized in the survey was in accordance with generally accepted standards of statistical procedure for this type of survey."[9]

In a 1959 case, the judge overruled the government's motion that the sample evidence be excluded on the grounds that it constituted hearsay evidence with the statement that the court was "satisfied that estimates, made on the basis of probability sampling . . . [were] sufficiently reliable to indicate the order of magnitude of income taxes that would be occasioned. . . ."[10]

[7] *Manual for Complex and Multidistrict Litigation*, Federal Judicial Center, 1973, sec. 2.712.

[8] Ibid.

[9] *State Wholesale Grocers v. The Great Atlantic and Pacific Tea Co.*, 154 F. Supp. 471, 488–499, *rev'd in part*, 258 F. 2d (CA 7 1958).

[10] *United States v. E.I. Du Pont de Nemours & Co.*, 177 F. Supp., 18–19 N.D. Ill. (1959).

In a recent case, in discussing the use of data from a sample as legal evidence, the judge commented that "sampling has long been considered an acceptable method of determining the characteristics of a large universe." He further observed that "such mathematical and statistical methods are well recognized by the courts as reliable and acceptable in determining adjudicative facts." He then cited numerous cases to support his view.

The judge also noted that "it was a principle recommended by the prestigious committee which wrote the *Manual for Complex and Multidistrict Litigation* that 'scientifically designed samples . . . meeting the tests of necessity and trustworthiness (be used to) contribute materially to shortening the trial of complex cases."

Finally the opinion states that "statisticians can tell us with some assurance what the reliability factors and probabilities are. Only the law can decide, as a matter of procedural and substantive policy, what probabilities will be required before the court will change the status quo by granting a remedy. Thus, in deciding whether a sample is adequate, practical limits to the fact-finding precision must be considered.[11]

Kecker points out that "courts are more likely, just as one would expect, to admit sample evidence gathered in conformity with accepted statistical standards than those produced by a professionally questionable technique."[12]

If there is any possibility that audit test data will be presented as evidence in a court of law, or if other samples of accounting data or records are to be so used, it follows that the sample design and execution must be impeccable and well documented. This point is emphasized by the manual for judges, which states that "the underlying data, methods of interpretation employed and conclusions reached in polls and samples should be made available to the opposing party far in advance of the trial."[13]

While it is inevitable that individual judges will deal with the admissibility of sample data as evidence in different ways, the precedents now exist and the trend is toward making such data acceptable in more and more cases, when the sample is properly designed and executed. In fact to an increasing degree, the judges seem to be "bending over backwards" to admit such evidence. This trend is attested to by the ruling of a judge with respect to excluding survey data which had been

[11] *Rosada v. Wyman*, 322 F. Supp. 1173 (1970).

[12] F. Kecker, op. cit., p. 116.

[13] *Manual for Complex and Multidistrict Litigation*, sec. 2.713.

criticized by an expert witness for improper design and execution of the sample investigation. The ruling admitted the data while declining to express any opinion about its probative values.[14]

The American Bar Association in one of its reports states:

> Samples involve the selection or examination of a part of the total units or elements about which information is needed . . . In order to project this report, however, the burden of proof rests upon the offeror to show that the sample was selected in accordance with accepted principles of sampling so that it properly represents the universe. Once this is established, the testimony of the statistical expert projecting the finding on the sample to the universe raises only questions of relevancy, materiality and weight. While such testimony should be excluded unless the judge is satisfied the sampling technique was reasonable, fair and adequate [the judge] might, in a non-jury case, admit the report on the sample itself if relevant, even if some doubt exists as to whether the projection is valid.[15]

In general, it must be stressed that where sample projections are subject to possible legal challenge, they must meet two tests: *necessity* and *validity*.

The test of necessity requires that a 100% examination of the records be either impossible or impractical. This requirement need not restrict the use of sample data as evidence to situations where very large numbers of records are involved. Where it can be shown that a complete (100%) audit of the records would be too costly or time-consuming, the sampling approach is defensible. Often it may be possible to demonstrate, even for fairly small populations, that the cost of a 100% examination would be excessive. In some cases, it would be possible to show that the cost of a complete audit would be more than the amount which might be recovered by the legal action. In other cases the cost, while not exceeding the recoverable amount, may be so considerable as to render such an approach impractical.

The Federal Judicial Council's manual states that "proof of necessity does not require a showing of total inaccessibility to proof of facts, but the offeror must show the impracticality of making his proof by conventional methods."[16]

The test of validity requires that the sampling plan be in accordance with generally accepted statistical principles and that the sampling be

[14] *United States v. Columbia Pictures Corp.*, 25 F.R.D. 497 (1960).

[15] "Streamlining the Big Case," The Report of the Special Committee of the Section on Antitrust Law, American Bar Association, 1958, pp. 32–33.

[16] *Manual for Complex and Multidistrict Litigation*, sec. 2.712.

executed in accordance with that plan. This proof may require the testimony of a witness who can qualify as an expert in the field of probability sampling. In addition, the auditors who executed the sample audit must be prepared to defend their audit determinations for each sampling unit examined, if such determinations are challenged. Indeed, this is the most vulnerable aspect of the sample projection. If it can be shown that there was inaccuracy or carelessness in the audit examination of the sampling units (nonsampling errors), no matter how sound the sample design, the projection is obviously subject to challenge as incorrect.

Of course, if it is shown that there are nonsampling errors, the projection may be amended, but the damage will have been done to the extent that questions will have been raised about the trustworthiness of the execution of the sampling process.

It may be contended in proceedings before a court or commission that the sample size used, no matter how large, was too small. Such a contention has no validity if an interval estimate has been used or if the sampling error was sufficiently small. It has been pointed out in prior chapters that the point estimate may be used for evidence only if the sampling error is innocuous. However, if the sample is not large enough to provide a trivial sampling error, the confidence limits of the interval estimate may be used.

While this point has been discussed extensively in Chapter 3, it might be noted that if the confidence limits are to be used, the limit selected on which to base a claim must be conservative. Thus, if a claim is made, based on a sample, for the collection of a sum of money, the claimant must specify the *lower* limit of his confidence interval estimate in order to have a defensible figure. It can be said with a high degree of assurance that *at least* this amount is due. The actual amount due is probably greater than that figure but cannot be determined.

An interval estimate has associated with it a specified confidence level. If such estimates are to serve as legal evidence, a sufficiently high confidence level must be chosen for them.

While there is some risk in sampling, there is also a risk of some kind (errors of measurement, clerical errors, errors in arithmetic, etc.) even in a 100% examination. Exact total values cannot actually be achieved even in a 100% examination because of the certain intrusion of clerical and other errors, especially in difficult audits.

For instance, in an audit of welfare payments by a government agency, there may be difficult and even controversial determinations required as to the eligibility of, and the correctness of payments to, some individuals. Similar problems may arise in dealing with contracts, which may be subject to varied interpretations by different individuals.

While the precise confidence level to be used may be a matter of opinion, it is the author's belief that confidence levels of 90% or higher, which are customarily used in audit tests, would be appropriate for legal evidence. In many fields where sampling is used, including the physical and social sciences, the 95% confidence level is most widely used. In *Rosada v. Wyman,* mentioned above, a 95% confidence level was apparently accepted by the court.

A well-designed and well-executed probability sample, supported by the testimony of an expert witness and demonstrated to be necessary, should have little difficulty in being sustained as appropriate legal evidence.

Chapter 10

Audit Risks Versus Statistical Risks

Any projection of sample results to a population is subject to risks. Thus, audit conclusions based on sample data are subject to such risks. These risks exist regardless of the method of sampling and apply whether probability sampling or judgment sampling has been used. The risks involved in a projection from a judgment sample cannot be evaluated on a probability basis.

There are two kinds of risks, audit risks and statistical risks. They are not identical, although they are related. There has been confusion in this area because some writers have equated the two types of risks.

The audit risks relate to the risk of an incorrect *audit* conclusion. The statistical risks relate to the risk of an incorrect conclusion about the parameters of the population.

AUDIT RISKS AND STATISTICAL RISKS

The audit risks arising from an audit test, based on a sample, vary with the test objectives. Not only do the test objectives vary within an audit but they are very different for different types of audits and auditors.

THE INDEPENDENT PUBLIC ACCOUNTANT AND AUDIT RISKS

Since the audit test provides only part of the information on which the public accountant's certifications are based, the audit test alone has only a limited impact on the audit conclusion.

The audit conducted by the independent public accountant is primarily concerned with the possibility of a *material* error in the financial statements of the company being examined. There are three general approaches used to provide the needed protection for the audit opinion:

1. The substantive tests directed towards establishing the reasonableness of the account balance;
2. The compliance tests directed toward detecting an excessive rate of errors or failures in the internal control system; and
3. The discovery test directed toward finding an instance of basic failure of the system, deliberate manipulation, or fraud.

In a compliance test, if failure to adhere to the internal controls is not revealed, the auditor will be ignorant of the need to enlarge the examination to offset the possible failure of the system to provide protection against a material error. Nevertheless, other audit work performed in the course of the examination may establish the possible existence of a material error.

Further, excessive deviations from the internal control system do not necessarily mean that there actually is a material error in an account balance or the financial statements. These are procedural errors and may not produce monetary errors. The repeated failure to obtain required authorizing signatures does not of necessity mean that fraud or a serious dollar value error exists. Nevertheless, the auditor would want to establish the existence of this state of affairs for his or her own protection and would want to caution management about the potential dangers involved.

The opposite risk in the compliance test is that the auditor may conclude that there is an excessive rate of deviation from the internal control system when in fact there is not. Perhaps there are actually a limited number of failures, possibly due to carelessness or rare haphazard circumstances. Of course, if the examination produces evidence of deliberate errors or manipulations, regardless of their number, the auditor will take further action. However, when actually there are only a few deviations, and those are not deliberate and of no monetary consequence, the improper conclusion that such errors are occurring at an excessive rate produces a "false alarm." The result of such a false conclusion is unnecessarily extended audit work and unnecessarily increased cost.

In general, however, while both of these types of errors—failure to detect excessive noncompliance or falsely claiming such noncompliance—are serious, they do not of necessity lead to a complete failure of the audit examination because of the protection afforded by the audit procedures. Nevertheless, such failures may have serious consequences.

In a substantive test (dollar value or variables sampling), performed to establish the reasonableness of a book record value, such as that for an inventory, accounts receivable, etc., failures like the above may be more serious.

The substantive test may fail to disclose that the investigated book value is materially misstated or may lead to the conclusion that the value is misstated when in fact it is not. Of course, other portions of the audit examination may cast doubts on the validity of a misstated value, but this does not provide assurance of a material misstatement or establish its amount. The determination of the amount of the misstatement must be established from sample evidence.

Thus, the impact of the audit risks in this situation is that the auditor may accept as reasonable an account balance which in fact is materially misstated and thus improperly certify the financial statements of the company audited. If it is falsely concluded from the test that the account balance is materially misstated when it is not, the result may be an improper adjustment to a book record value or a 100% check which proves the auditor's conclusion to be wrong.

Finally, there is an entirely different audit risk in audit tests of all kinds. There may be a failure to include in the sample, and thus a failure to detect, sampling units which show such important deviations as failure to comply with proper procedures or generally accepted accounting methods, or which provide evidence of massive errors or fraud.

RISKS AND THE SUBSTANTIVE TEST

The audit risks involved in the substantive test concern the possibility that the determination from the sample may lead to the failure to find a *material* error when one exists *or* to the incorrect conclusion that a material error exists when in fact there is none.

It should be noted that the consequences of these erroneous conclusions in an audit test are by no means of equal importance. A failure to disclose a material error in the financial statement, when one exists, can be catastrophic to the auditor. Such a failure may cause litigation and damage or even ruin the auditor's reputation. On the other hand, claiming a material error where none exists is likely to cause the auditor to extend the audit, and may even, in some instances, result in a 100% investigation. While this unnecessary extension of audit activity may result in increased audit costs and may damage the individual client's view of the auditor's efficiency, it is by no means as damaging as the other type of failure.

However, if adequate additional investigation is undertaken, it is not likely that the auditor will be misled by the single test and improperly qualify or refuse a certification. The net result then is unnecessary additional audit cost.

At this point, it is important to emphasize that the amount of error which is considered to be material cannot be defined precisely so that all auditors would arrive at the same figure. Its definition is largely a matter of judgment. *SAS 1* states "that the auditor's decision as to the monetary amount or frequency of errors that would be considered material should be based on his judgment in the circumstances in the particular case." [1]

While there has been much discussion about the concept of materiality, no uniform definition has been generally adopted nor has any single method been established for precisely fixing the amount in each situation. It is not likely that a definitive method of fixing a single value that would be accepted by all auditors will ever be found.

Thus, the objective of the substantive test is to establish whether an account balance is misstated by at least an amount which can only be fixed approximately.

The method of sampling appropriate in substantive tests is estimation sampling for variables, sometimes referred to as dollar value sampling.[2] In this method an interval estimate of the audited dollar value or total amount of error in the book record value is obtained.

The *statistical* risks in this situation are that the interval estimate of the audited value will include the book record value when it should not or that it will fail to include the book record value when it should. The audit consequences of these errors are that in the first case an error will go undetected and in the second an error will be indicated where none exists.[3]

Again, it must be recognized that there is a difference between the audit risks and the statistical risks. The audit risks relate primarily to the failure to disclose a material error when it exists and secondarily to the claim of such an error when it does not exist.

The statistical risk relates to the probability that the projected confidence interval does not contain the results which would have been achieved by a 100% examination or that it contains that value when it should not.

[1] *Statement on Auditing Standards 1*, p. 39.

[2] Not to be confused with Dollar *Unit* sampling as described in Chapter 7.

[3] While such an explanation seems to imply that these risks are the same as those referred to as type 1 and type 2 errors in statistics textbooks, because of the similarity of wording, as will be seen this is an apparent and not a real similarity.

While under some circumstances the two risks are identical, in many they are not. In the extreme case, assume that there are no errors at all in the accounting system. Under such circumstances no matter what the statistical risks are, there is *no* risk of failing to find a material error and thus of failing to disclose that financial statement is materially misstated.

On the other hand, in some statistical projections in the audit situation where there is no material error, there is both a statistical *and* audit risk of implying that one does exist. For instance, if an inventory value is projected from a sample to establish the reasonableness of a book record value, there is a risk that the projected confidence interval will not contain the population (100% sample) value. This risk is equal to 100% minus the confidence level.

However, if a *reasonable* sample precision is attained, there is protection against a false conclusion even in the 5% or 1% of the cases which would provide such an unfortunate projection. Although the auditor may falsely come to the conclusion that the book record value is wrong, the probability that the indicated error would be of a material amount is very small. And thus, since no action will be taken, the audit risk is virtually nonexistent.

A distinction must be made between these two types of risks, although it must still be recognized that they are interrelated. Thus, if material error does exist, there *is* a risk, *both* of a statistical and audit nature, of failing to detect the error.

Using the traditional approach of classical statistics, the method is developed by creating the null hypothesis that there is no difference and then attempting to disprove this null hypothesis by showing that the difference between the point estimate and the book record value is too great to be explained by the sampling error. This may be demonstrated by the fact that the book record value is not included within the confidence interval. When such a situation occurs it may be stated that there is a probability, at least equal to the confidence level, that the difference is *statistically significant*.

Thus, when the difference between the sample point estimate and the population value (in this case the book record value) exceeds a figure which can be accounted for by the sampling error, there probably is a real difference. In other words, the disparity cannot be said to be an accident of sampling.

However, a real difference need not be of any specific magnitude to be declared statistically significant. Even the smallest difference can be declared to be significant if the sample is large enough. Yet the auditor is interested in a difference only if it is material in the accounting sense.

It might be possible to overcome this difficulty if the tested difference between the point estimate and the book record value is *increased* by the minimum amount of the difference which would be considered to be material. But the amount considered to be material cannot be established exactly. It is an approximate figure and the question at hand is not whether there is *any* disparity after the inclusion of the material error amount, but rather how much of a difference exists. A very small difference beyond the approximate material amount certain would not be considered of as great an importance as a considerable amount beyond that point.

The real audit purpose is to establish the *amount* of the difference between the audited and book record value, so that the auditor can consider the impact of that difference on the financial statement. On the other hand no difference can be considered to be material unless it is demonstrated to be *statistically* significant.

A substantive test therefore has as its purpose an estimate of the magnitude of the difference, if any, between the book record value and the audited value of an account. When the auditor is aware of the minimum and maximum amount (confidence limits) of such a difference, he is then in a position to evaluate the result, providing the spread between these limits is not too great *and* the difference has been declared to be statistically significant.

With a very small sample, at a given confidence level, the spread between the upper and lower confidence limits may be so wide as to include the book record value even if a very large disparity actually exists. A large sample will limit the interval so that even if the disparity is small, the confidence limits will not include the book record value.

However, for a given sample size, the probability of detection of any disparity is related to the confidence *level* used. The higher the confidence level, for a given sample size, the wider the confidence interval. Thus, the use of a 95% confidence level will produce a wider confidence interval than will the use of a 90% confidence level and is less likely to confirm a disparity when it exists.

This should be apparent when it is considered that attesting to a disparity at a higher confidence level means that there is less risk attached to the conclusion that the disparity exists.

On the other hand, the increased spread of the confidence interval makes less probable the error of the opposite kind, namely the declaration that a disparity exists when in fact there is none.

If the book record value is included within the interval estimate (between the upper and lower limits) no matter how large a difference exists between the point estimate and the book record value, no difference of *any* magnitude can be claimed, since the difference may be an accident of sampling.

If the book record value is not included within the interval estimate, the difference between the point estimate and the book record value most probably is not an accident of sampling, but may not be of any consequence unless it is material in the accounting sense. This difference between the point estimate and book record value may not be considered to be material unless the book record value differs from the point estimate by at least a material amount.

The more important of the two possible errors from an audit viewpoint is the failure to detect a material error. However, any disparity cannot be declared material unless it is statistically significant. It is statistically significant only if the book record value is not included in the confidence interval (is not between the upper and lower confidence limits). The risk that the book record value will be contained within the confidence interval even though there is a real disparity between the book record value and the true or audited value is a function of the actual magnitude of the disparity, the sample size used in the audit sample, and the confidence level.

The confidence level, representing the assurance that a disparity between the book record value and the audited value actually exists when the book record value is outside of the confidence interval, controls the error of the second kind, the declaration of a disparity when none exists.

Enlarging the sample provides increased protection against a declaration of a disparity when none exists, but an increase in the spread between the upper and lower confidence limits increases the risk of the opposite kind of error, failure to detect a disparity when it actually exists (error of the first kind). The sample size, for a given confidence level, controls this more serious kind of error, failure to detect a disparity when it exists. As suggested in Chapter 3, the sample size should be sufficient to provide a sampling precision of $\frac{1}{2}M$ where M is the amount considered to be material.

The relationship discussed above may be seen in the following tables. In Table 10-1, the effect of various sample precisions actually achieved is given.

In this example, a material error equal to 1.5 times the minimum amount considered to be material actually exists in the book record value. For a fixed confidence level (95%) and for various sample precisions achieved by the sample used, the probability of failing to detect the occurrence of *any* error and the probability of failing to detect a material error is given in Table 10-1.

It is seen that there is a high probability of failing to detect a material error when the sampling error is as much as or more than the amount of the material error, even though the actual error is 50% greater than the minimum amount considered to be material.

It is to be remembered that the sample precision (sampling error) is determined by the sample size used for a given population and confidence level. This table shows the futility of using inadequate sample sizes.

Cost, time, and other considerations are often used to excuse the use of very small samples. But such small samples are likely to be useless in detecting material errors when they exist.

Table 10-2 illustrates the impact of the selected confidence level upon the failure to detect errors. Before considering this table it must be remembered that, for a given population, to achieve the same sample precision at a higher confidence level a disproportionately larger sample must be used.

It will be seen from Table 10-2 that the increase in the confidence level from 90% to 95% reduces the probabilities of failure to detect error but to a limited degree. As will be shown later, the confidence level is more closely related to the other kind of error, the probability of *falsely* declaring that an error or a material error exists.

This second kind of error results in a false conclusion, based on the sample projection, either that some error exists or that a material error exists. The probability that some error will be indicated although none exists is the complement of the confidence level (1 − CL). This risk is

Table 10-1 ***Probability of Failure to Detect Any Error and Material Error for Various Sample Precisions and Actual Error Equal to 1.5 Times the Minimum Error Amount Considered Material***
(95% Confidence Level)

Achieved Sample Precision*	Probability of Failure to Detect Material Error†
1.50 M	50.0%
1.00 M	16.0%
0.75 M	2.5%
0.50 M	Less than 0.001%

*Where M equals minimum amount of error considered to be material.

†The probability (one-sided) that the book record value will be contained within the confidence interval.

NOTE: The figures in this table are based on the assumption of a normal sampling distribution. See Chapter 3 for a discussion of the impact of a nonnormal sampling distribution. The relationships will be similar for nonnormal sampling distributions.

Table 10-2 ***Probability of Failure to Detect a Material Error for Various Confidence Levels When Sample is Equal to One-half Minimum Error Amount Considered to Be Material and Actual Error of Minimum Amount Considered Material***

Confidence Level, %	Probability of Failure to Detect Material Error*
90	57%
95	51%

*The probability (one-sided) that the book record value will be included within the confidence interval or that the difference between the book record value and the point estimate will be less than a material amount.

NOTE: The figures in this table are based on the assumption of a normal sampling distribution. See Chapter 3 for a discussion of the impact of the nonnormal sampling distribution. The relationship will be similar for nonnormal sampling distributions. Standard Error = SE/*t* or sampling error divided by *t*.

entirely a function of the confidence level and is controlled by its selection. However, the confidence level does not exclusively control the probability that a material error will be falsely indicated by the sample projection when the following rules are applied. The rules are that the book record value must be outside the confidence limits *and* that the difference between the point estimate of the audited value and the book record value must be at least equal to the minimum error amount considered to be material.

These relationships are shown in Table 10-3. The probability of false indication of a material error where there is none is a function of the sample precision compared to the amount of the material error. It is small when the sample precision achieved is less than the amount specified as a material error.

RISKS AND THE COMPLIANCE TEST

While the audit risks involved in a compliance test are similar to those encountered in the substantive test, nevertheless there are some important differences.

It must be remembered that the compliance test relates to an evaluation of the operation of the system, not a specific dollar value. Its function is to determine whether there is an excessive rate of error or of

Table 10-3 ***Probability of Falsely Declaring a Material Error Exists, When Actual Error Equals One-half of Material Amount, Based on a Sample Projection for Various Sample Precisions***
(95% Confidence Level)

Sample Precision*	Probability of Falsely Concluding that a Material Error Exists†
1.50*M*	26%
1.00*M*	16%
.50*M*	2½%

**M* is equal to the minimum error amount considered to be material.
†The probability (one-sided) that the confidence interval does not include the book record value *and* the book record value is further from the point estimate than the material amount.
NOTE: The figures in this table are based on the assumption of a normal sampling distribution. See Chapter 3 for a discussion of the impact of the nonnormal sampling distribution. The relationship will be similar for a nonnormal sampling distribution.

failure to comply with the internal control system. Keep in mind the fact that inadvertent failures to comply are not rare occurrences in most large accounting systems. An exception to this willingness to accept such events when they are not frequent exists only for errors of the most critical types.

Again, the audit risks involved are failing to detect such an excessive rate of error, when it exists, or claiming that such excessive deviations occur when in fact they do not.

In considering the consequences of such failures, it must be remembered that the compliance test is primarily designed to establish what further audit efforts may be necessary because of the possible failure of the protection provided by the internal control system.

SAS 1 defines the second standard of field work as follows:

> There is to be a proper study and evaluation of the existing internal control as a basis for reliance thereon and for the determination of the resulting extent of the tests to which auditing procedures are to be restricted.[4]

[4]*Statement on Auditing Standards 1*, p. 13.

The "study and evaluation" of the internal control system is then described as a qualitative evaluation to establish the extent of the protection provided by the system *and* a test to establish the frequency of failures to comply with the system.

The qualitative examination of the protective features of the internal control system is subjective in nature and different auditors may arrive at different conclusions about its protective ability.

Further, there may be frequent failures to comply with internal controls without a resulting material error. For instance, there may have been frequent careless failures to secure proper authorizing signatures. While this is a dangerous situation, it will not necessarily result in any improper accounting values.

Bearing this in mind, it will be seen that from a public accountant's viewpoint, failure to detect an excessive rate of error may *or* may not result in a failure to establish the existence of a material error in the financial statement. However, since failure to detect an excessive rate of deviation may result in less extensive other testing and auditing procedures, the probability of detecting a material error when it exists is therefore further reduced.

However, unless the auditor relies exclusively on the results of the compliance test together with the qualitative evaluation of the internal control system and performs *no* other tests or audit evaluations, the likelihood of a failure to detect a materially misstated financial statement will be much reduced by other audit steps. It is unlikely that the auditor will do nothing but perform a test and evalution of the internal control system.[5]

On the other hand, the governmental auditor or internal auditor, whose mission is to detect failures to comply with regulations and prescribed procedures and to make appropriate audit findings, is different. Since in most cases these auditors do not certify the overall financial statement as do public accountants, their compliance test stands alone. Here the failure of the compliance test is a failure of their audit mission.

Contrariwise, a test resulting in an indication of an excessive error rate when it does not exist (a false alarm) may result in an unnecessary and costly widening of the extent and scope of the audit examination.

Once again, however, it must be remembered, in considering the statistical as well as the audit risks, that the rate of error considered to be

[5] *Statement on Auditing Standards 1* indicates that it is not contemplated "that the auditor will place complete reliance on internal control to the exclusion of other auditing procedures with respect to material amounts in the financial statement," para. 320.71.

excessive (MTER) in a compliance test cannot be determined exactly and will vary from auditor to auditor in the same situation and even for the same type of error. Of necessity, this is a judgment determination.

Thus, from a statistical viewpoint, the objective of a compliance test (estimation sampling for attributes) is to establish whether the rate of some type of failure (or error) in the population is more or less than a rate (the MTER) which can only be fixed approximately. In fact, the MTER should not be looked upon as a precise figure but a general area or zone of error.

This objective is accomplished by providing an interval estimate of the rate of occurrence of error in the population and comparing the MTER to the limits of the confidence interval estimate.

As explained in Chapter 2, there are three possible decisions. If the MTER is in excess of the upper confidence limit the rate of occurrence in the population is considered to be acceptable. If it is below the lower limit, the rate in the population is concluded to be excessive. If the MTER falls between the confidence limits, the result of the test is indeterminate.

The *statistical* risk involved in this situation is that the interval estimate of rate of occurrence in the population will include the MTER when it should not or that it will not include the MTER when it should. These failures are related to the more usually considered error that the confidence interval may not include the population value but are not entirely dependent on that error.

The statistical risks in the compliance test are related to the decision to use a one-sided or two-sided interval estimate method (see Chapter 2).

If a one-sided upper confidence limit estimate is used, as suggested in AICPA literature, only two decisions can be made. If the MTER is above the upper confidence limit, the error rate (if any) is deemed acceptable. If the MTER is below the upper limit, the error rate is considered unacceptable.

When the AICPA method is used, their literature suggests that the sample size be determined by selecting a sample such that if some "expected" rate of occurrence is found in the sample the resulting upper limit projected from the sample will be equal to the MTER.

If the expected number or a smaller number of errors is found in the sample, it is deemed acceptable. The probability that an incorrect conclusion will be reached that the occurrence rate is acceptable when it is not then is determined by the selected confidence level. The risk is equal to 100% minus the confidence level This is the probability that a population containing an error rate in excess of the MTER will nevertheless produce a sample containing fewer than the expected number of errors.

THE GOVERNMENTAL AUDITOR AND AUDIT RISKS

Since, as previously observed, the governmental auditor usually issues no overall certification of financial statements, the audit risks for him are usually related to *each* finding in the report.

To the extent that the audit finding is based on the discovery of an instance of a given type of failure in the sample, there is no risk involved in claiming that such a condition actually exists. An example is at hand in the sample. However, deciding whether the discovery of one or a few errors in the sample indicates a condition worthy of some action, since if it is widespread, it does involve some risks. In many government audits some errors are discovered; the question to be resolved is whether such errors are sufficiently pervasive as to require action. An outstanding example of this situation occurs in audits of welfare payments. Often payments to ineligibles or inaccurate payments are frequent. Further, there is probably some irreducible level of such errors.

Once again, there may be a failure to disclose a bad situation (excessive rate of failures) or a false claim that a very bad situation exists when it does not. However, these determinations are usually achieved on the basis of a single test (sample).

The consequences of these risks for governmental auditors are different from those for independent public accountants in the case of the compliance test. The result of a failure to disclose a bad situation will be that the condition will continue to exist, possibly involving large unnecessary costs, improper payments, and even fraud. A false claim of a bad situation may result in excessively costly action to correct a minor situation.

Further, as previously discussed, the governmental auditor may use a sample to establish or check upon a total dollar value. The auditor may want to establish disallowances in audits of cost-plus contracts, of claimed deductions in a business tax return, of improper expenditures by a governmental unit, etc. The intent in each case is to collect the amount from the audited organization.

An audit failure in this situation can mean a failure to collect the proper amount due or an attempt to collect an improper amount which can be successfully challenged.

THE CONFIDENCE LEVEL PROBLEM

The selection of a confidence level, for use in an audit test based on a probability sample, has long been considered one of the most difficult and controversial aspects of the use of statistical sampling techniques in auditing.

Since the confidence level must be selected by the individual auditor and is to at least some extent judgmental, it is important that its meaning and significance be fully understood.

SAS 1 states that the "specification of the precision and reliability necessary in a given test is an auditing function and must be based upon judgment in the same way as the decision as to audit satisfaction required when statistical sampling is not used."[6]

While such a selection is a subjective determination, it is required that the confidence level be included with probability sampling results, so that others who might disagree with the assurance provided can recalculate on the basis of their own selection of a confidence level.

MEANING AND SIGNIFICANCE OF THE CONFIDENCE LEVEL

All statistical sample evaluations have associated with them a confidence or probability level.

Thus, an interval estimate of a dollar value achieved through the use of estimation sampling for variables has associated with it a confidence level expressive of the risk involved in relying on that result.

The purpose of an interval estimate obtained from a probability sample is to provide an interval within which there is a stated probability (confidence level) that the value (average, total, percent, etc.) computed from a 100% sample would be included. To express this conclusion in a more traditional fashion, if sampling of the type and size used were repeated innumerable times, in a percentage of such trials equivalent to the confidence level, the population value which would have been obtained from a complete audit will be contained in that range.

Thus, a statement of an interval estimate together with a confidence level of, say, 95% means that there is a 5% *risk* that such an interval will not include the result obtainable from a 100% sample. In other words, if the same test, using probability sampling, were repeated 100 times, it is probable that on five such tests, the results obtainable from a 100% sample would not be included in the established confidence interval.

[6]*Statement on Auditing Standards 1*, para. 320.A.03. It is to be remembered that the AICPA publications use the term "reliability" rather than "confidence" level.

The nature of such a risk must be examined for its impact on the audit test. Assume that an auditor wishes to establish the reasonableness of a stated inventory book record value of $1,216,374. The auditor takes a statistical sample using estimation sampling for variables techniques and projects an interval estimate for the total inventory value of from $1,120,619 to $1,160,619 at a 95% confidence level.[7]

On the basis of this result, the auditor can conclude that the book record value is not correctly stated. Of course, it is assumed that the sampling operation would be rechecked and an attempt made to establish the reasons for such a misstatement of the inventory value. Such reasons can often be deduced from the type of errors found in the sample.

However, the auditor does not have *absolute* assurance of a misstated inventory value, since some risk is involved in any sampling. If the auditor is satisfied that the assurance is sufficiently high (risk sufficiently small), the interval estimate can be relied on.

On the other hand, it is not generally recognized that there is a relation between the level of assurance achieved and the magnitude of the possible error, if the sample used is one of those few (in this case 5% or 1 in 20 of the possible samples) which produces an interval which does *not* include the result obtainable from a 100% sample.

It will be remembered that the interval estimate is achieved for variables samples by adding and subtracting a value (the sampling error) from the point estimate. The sampling error is essentially t times the standard error (standard deviation of the sampling distribution). The confidence level achieved is a function of the multiplier (t) used in the calculation. For the 95% confidence level, the multiplier is 1.96.

If a multiplier of 3.3 had been used, a 99.9% confidence level would have been achieved. With such a high confidence level, there would be virtual certainty that the 100% sample result would have been included in the interval estimate. The chance that it would not be is only 1 in 1000.

If the auditor who used the 95% confidence level should be so unfortunate as to draw a sample which is actually one of those few (1 in 20) which produce an interval which does not contain the population value, the auditor will not be off by an indefinite amount and almost certainly not by an amount more than the difference between the limit of that estimate and the limit from a 99.9% confidence interval.

Since there is an extremely small probability (1 chance in 1000) of the population value not being contained within the confidence interval for

[7]A similar discussion can be developed using one-sided confidence limits, but since the two-sided estimate serves better as an example, that method is used here.

the 99.9% confidence level, it may for all practical purposes be assumed that such a departure will not happen. Therefore, the sampling error based on a 99.9% confidence level may be considered to be the "*maximum*" sampling error or the maximum disparity between the sample point estimate and the population value that might arise due to sampling fluctuations.

Another examination of the above result provides a further interpretation. It is apparent that the extent of the misstatement of the difference between the book record value and the confidence limit of the estimate, in the case of the rare sample which produces a confidence interval not containing the population value, is a function of the confidence *level* used.

For instance, in the example cited above the confidence interval at the 95% confidence level was $1,120,619 to $1,160,619 as contrasted with a book record value of $1,216,374. If the upper confidence limit is used to establish the overstatement, an overstatement of at least $55,755 results.

Based on a 99% confidence limit, the overstatement on the same basis would be $42,082. The difference between these two estimates of overstatement is $13,673. If a 90% confidence level had been used, the disparity would have been $58,919!

A comparison between the sampling error and the sampling error of an unusual sample (at the 99.9% confidence level) for various confidence level estimates can be seen in the table below.

Sampling Errors, Variables Estimation Samples, Assuming a Standard Error of the Arithmetic Mean* Equal to $10 (Infinite Population)

Confidence Level Used (Two-Sided) %	Indicated Sampling Error	Maximum Sampling Error	Maximum Difference for Unusual Sample	
			Amount	Percent
80	$12.80	$33.00	$20.20	158
90	16.50	33.00	16.50	100
95	19.60	33.00	13.4	68
99	25.80	33.00	7.2	25

*The sampling error of this arithmetic mean for an infinite population is merely *t* times the standard error.

It is quite evident that not only is there a higher risk of not including the population value within the bounds of the confidence interval for low confidence levels but the maximum disparity between the point estimate produced by the sample and the population value will be greater as well.

On the other hand, it must be emphasized that the probability of a disparity as large as that indicated as the "maximum" sampling error is very unlikely and that the actual disparity will most probably be much smaller.

Nevertheless, the two-fold impact resulting from the use of low confidence levels must be considered.

In the example cited above, the point estimate from the sample of the inventory value was $1,140,619 with a sampling error of ±$20,000 at the 95% confidence level. This can be interpreted to mean that the auditor is 95% certain that there is an overstatement of between $55,755 and $95,755 in the inventory value. These figures are the difference between the two confidence limits and the book record value. However, he can be virtually certain that, even if he is unfortunate in the sample he selects, he has a sampling error of not more than $33,673 with a 99.9% confidence level, and that the overstatement is between $42,082 and $109,328.[8]

[8]If the sampling error is $20,000 at the 95% confidence level, the standard error is $20,000 ÷ 1.96 or $10,204. This figure times 3.3 equals $33,673.

Appendix A

Two-Step Sampling Tables

Two-step sampling involves the use of an initial sample (n_1) which is audited. If more than a specified number of errors are found in the first sample, an additional sample (n_2) is taken. The confidence interval for the resulting combined initial and incremental sample can then be found in the following tables.

The sampling plan on which these limited tables are based involves an initial sample ranging in size from 50 to 100. If *no* errors are found in the sample, sampling is discontinued and the upper limit of the confidence interval estimate can be obtained from Table A-1.

If errors are found in the initial sample, additional sampling units may be selected and audited. By reference to Table A-2 the two-sided confidence interval estimate can be established for a selected confidence level.

Interval estimates are provided for additional sample sizes of 60, 120, and 180. The number of occurrences in the sample specified in the table refers to the total number of occurrences in *both* the initial and additional samples.

The use of these tables may be illustrated by a situation in which an auditor is concerned with a compliance test for a specified type of deviation. The auditor established the MTER as 5% and wished to use a 95% confidence level. In accordance with the procedure outlined in Chapter 2, pages 22 to 26, an initial sample of 60 is selected because if *no* errors are found in the initial sample, the upper limit of the confidence interval estimate will be 4.9%. This would indicate an acceptable situation.

If any errors are found in the first sample, the auditor will audit an additional sample. An additional sample has to be taken since four errors are found in the first sample. The decision is to expand the sample by selecting an additional sample of 120. The auditor finds 12 additional occurrences in the additional sample of 120, for a total of 16. Table A-2 (p. 197) indicates that 16 occurrences in the sample of 60 plus 120 provide a confidence interval estimate of 5.2 to 14.0% at the 95% confidence level. This indicates an unacceptable condition since the rate of occurrence in the population is at least 5.2% or at least as much as the MTER (5%).

A more detailed example is provided in Chapter 2, pages 26 to 34.

Table A-1 ***Upper Confidence Limits, Single Sample, with Zero Occurrences in Sample, at Various Confidence Levels*** ***(N = 1000)***

Sample Size	Upper Confidence Limit		
	90%	95%	99%
50	4.4%	5.7%	8.6%
60	3.7	4.7	7.2
70	3.1	4.0	6.1
80	2.7	3.5	5.4
90	2.4	3.1	4.8
100	2.2	2.8	4.3

NOTE: Since the sample sizes in the above table are very small compared to the population of 10,000 or more, the population size has a very limited effect on the upper limits. Therefore, the section of the table for N = 10,000 can be used for populations of 10,000 and over.

Table A-2
Two-Step Sampling Confidence Limits*
Initial Sample **(n_1) = *50***
Confidence Level = 90% (Two-Sided), 95% (One-Sided)
N = *1000*

Occurrences in Sample	Single Sample n_1		Two-Step (Double) Sample ($n_1 + n_2$)					
			n_2 = 60		n_2 = 120		n_2 = 180	
	LL	UL	LL	UL	LL	UL	LL	UL
0	0.0%	4.4%	—	—	—	—	—	—
1			0.1%	5.8%	0.1%	5.7%	0.1%	5.7%
2			0.4	6.2	0.3	5.7	0.2	5.7
3			0.9	7.1	0.6	5.8	0.5	5.7
4			1.4	8.1	1.0	6.0	0.8	5.7
5			1.9	9.2	1.3	6.5	1.0	5.7
6			2.5	10.3	1.7	7.0	1.3	5.9
7			3.2	11.4	2.1	7.6	1.6	6.1
8			3.8	12.4	2.5	8.2	1.9	6.4
9			4.5	13.5	3.0	8.8	2.3	6.8
10			5.2	14.6	3.4	9.5	2.6	7.2
11			5.9	15.6	3.9	10.2	2.9	7.7
12			6.6	16.7	4.3	10.9	3.3	8.1
13			7.4	17.7	4.8	11.5	3.6	8.6
14			8.1	18.8	5.3	12.2	4.0	9.1
15			8.8	19.9	5.8	12.9	4.3	9.5
16			9.6	20.9	6.3	13.6	4.7	10.0
17			10.4	21.9	6.7	14.2	5.0	10.5
18			11.1	22.9	7.2	14.9	5.4	11.0
19			11.9	23.9	7.7	15.6	5.8	11.5
20			12.7	24.9	8.2	16.2	6.1	12.0
21			13.8	25.9	8.7	16.9	6.5	12.4
22			14.2	26.9	9.2	17.5	6.9	12.9
23			15.0	27.9	9.7	18.2	7.3	13.4
24			15.8	28.9	10.3	18.8	7.6	13.9
25			16.6	29.9	10.8	19.5	8.0	14.4

*The two-step (double) sample confidence limits in this table are based on an initial sample of size n_1 with an additional sample of size n_2 *only* if occurrences are found in the initial sample.

Table A-2
Two-Step Sampling Confidence Limits*
Initial Sample (n_1) ***= 50***
Confidence Level = 90% (Two-Sided), 95% (One-Sided)
N = ***5000***

Occurrences in Sample	Single Sample n_1		Two-Step (Double) Sample ($n_1 + n_2$)					
			$n_2 = 60$		$n_2 = 120$		$n_2 = 180$	
	LL	UL	LL	UL	LL	UL	LL	UL
0	0.0%	4.5%	—	—	—	—	—	—
1			0.1%	5.9%	0.1%	5.8%	0.1%	5.8%
2			0.4	5.4	0.3	5.8	0.3	5.8
3			0.8	7.3	0.6	6.0	0.5	5.8
4			1.3	8.3	0.9	6.2	0.7	5.8
5			1.8	7.4	1.2	6.7	1.0	5.9
6			2.4	10.5	1.6	7.2	1.2	6.1
7			3.1	11.6	2.0	7.8	1.5	6.4
8			3.7	12.7	2.4	8.5	1.8	6.7
9			4.4	13.8	2.8	9.1	2.1	7.1
10			5.1	14.9	3.3	9.8	2.4	7.5
11			5.7	15.9	3.7	10.5	2.8	8.0
12			6.5	16.9	4.2	11.2	3.1	8.4
13			7.2	18.0	4.6	11.8	3.4	8.9
14			7.9	19.0	5.1	12.5	3.8	9.4
15			8.6	20.1	5.6	13.2	4.1	9.9
16			9.4	21.2	6.0	13.9	4.5	10.4
17			10.1	22.2	6.5	14.5	4.8	10.8
18			10.9	23.2	7.0	15.2	5.2	11.3
19			11.7	24.2	7.5	15.9	5.5	11.8
20			12.4	25.2	8.0	16.6	5.9	12.3
21			13.2	26.2	8.5	17.2	6.3	12.8
22			14.0	27.2	9.0	17.9	6.6	13.3
23			14.8	28.2	9.5	18.5	7.0	13.8
24			15.6	29.2	10.0	19.2	7.4	14.3
25			16.4	30.2	10.5	19.8	7.7	14.8

*The two-step (double) sample confidence limits in this table are based on an initial sample of size n_1 with an additional sample of size n_2 *only* if occurrences are found in the initial sample.

Table A-2
*Two-Step Sampling Confidence Limits**
Initial Sample (n_1) = *50*
Confidence Level = 90% (Two-Sided), 95% (One-Sided)
N = *10,000*

Occurrences in Sample	Single Sample n_1		Two-Step (Double) Sample ($n_1 + n_2$)					
			n_2 = 60		n_2 = 120		n_2 = 180	
	LL	UL	LL	UL	LL	UL	LL	UL
0	0.0%	4.5%	—	—	—	—	—	—
1			0.1%	5.9%	0.1%	5.8%	0.1%	5.8%
2			0.4	6.5	0.3	5.8	0.3	5.8
3			0.8	7.1	0.6	6.0	0.5	5.8
4			1.3	8.3	0.9	6.3	0.7	5.8
5			1.8	9.4	1.2	6.7	0.9	5.9
6			2.4	10.5	1.6	7.3	1.2	6.1
7			3.0	11.6	2.0	7.9	1.5	6.4
8			3.7	12.7	2.4	8.5	1.8	6.7
9			4.4	13.8	2.8	9.2	2.1	7.1
10			5.0	14.9	3.3	9.8	2.4	7.6
11			5.7	15.9	3.7	10.5	2.7	8.0
12			6.4	17.0	4.1	11.2	3.1	8.5
13			7.2	18.0	4.6	11.9	3.4	8.9
14			7.9	19.1	5.1	12.6	3.8	9.4
15			8.6	20.2	5.5	13.2	4.1	9.9
16			9.4	21.2	6.0	13.9	4.4	10.4
17			10.1	22.2	6.5	14.6	4.8	10.9
18			10.9	23.3	7.0	15.3	5.1	11.4
19			11.6	24.3	7.5	15.9	5.5	11.9
20			12.4	25.3	8.0	16.6	5.9	12.4
21			13.2	26.3	8.5	17.3	6.2	12.9
22			14.0	27.3	9.0	17.9	6.6	13.3
23			14.8	28.3	9.5	18.6	7.0	13.8
24			15.6	29.3	10.0	19.2	7.3	14.3
25			16.4	30.3	10.5	19.9	7.7	14.8

*The two-step (double) sample confidence limits in this table are based on an initial sample of size n_1 with an additional sample of size n_2 *only* if occurrences are found in the initial sample.

Table A-2
Two-Step Sampling Confidence Limits*
Initial Sample (n_1) ***= 60***
Confidence Level = 90% (Two-Sided), 95% (One-Sided)
N = ***1000***

Occurrences in Sample	Single Sample n_1		Two-Step (Double) Sample ($n_1 + n_2$)					
			$n_2 = 60$		$n_2 = 120$		$n_2 = 180$	
	LL	UL	LL	UL	LL	UL	LL	UL
0	0.0%	3.7%	—	—	—	—	—	—
1			0.1%	4.9%	0.1%	4.7%	0.1%	4.7%
2			0.4	5.5	0.2	4.8	0.2	4.7
3			0.8	6.3	0.6	5.0	0.5	4.7
4			1.2	7.3	0.9	5.3	0.7	4.8
5			1.8	8.3	1.2	5.8	1.0	4.9
6			2.3	9.4	1.6	6.4	1.3	5.2
7			2.9	10.4	2.0	7.0	1.5	5.5
8			3.5	11.4	2.4	7.6	1.9	5.9
9			4.1	12.4	2.8	8.3	2.2	6.3
10			4.8	13.4	3.2	8.9	2.5	6.8
11			5.4	14.4	3.7	9.6	2.8	7.2
12			6.1	15.4	4.1	10.2	3.1	7.7
13			6.7	16.3	4.6	10.9	3.5	8.1
14			7.4	17.3	5.0	11.5	3.8	8.6
15			8.1	18.2	5.5	12.2	4.1	9.1
16			8.8	19.2	5.9	12.8	4.5	9.6
17			9.5	20.1	6.4	13.4	4.8	10.0
18			10.2	21.0	6.8	14.1	5.2	10.5
19			10.9	22.0	7.3	14.7	5.5	11.0
20			11.6	22.9	7.8	15.3	5.9	11.4
21			12.3	23.8	8.3	15.9	6.3	11.9
22			13.1	24.7	8.7	16.6	6.6	12.4
23			13.8	25.6	9.2	17.2	7.0	12.8
24			14.5	26.5	9.7	17.8	7.3	13.3
25			15.2	27.4	10.2	18.4	7.7	13.8

*The two-step (double) sample confidence limits in this table are based on an initial sample of size n_1 with an additional sample of size n_2 *only* if occurrences are found in the initial sample.

Table A-2
Two-Step Sampling Confidence Limits*
Initial Sample *(n_1) = 60*
Confidence Level = 90% (Two-Sided), 95% (One-Sided)
N = ***5000***

Occurrences in Sample	Single Sample n_1		Two-Step (Double) Sample ($n_1 + n_2$)					
			n_2 = 60		n_2 = 120		n_2 = 180	
	LL	UL	LL	UL	LL	UL	LL	UL
0	0.0%	3.8%	—	—	—	—	—	—
1			0.1%	5.0%	0.1%	4.9%	0.1%	4.8%
2			0.3	5.7	0.3	4.9	0.2	4.8
3			0.7	6.5	0.5	5.2	0.4	4.9
4			1.2	7.5	0.8	5.6	0.7	5.0
5			1.7	8.6	1.2	6.1	0.9	5.2
6			2.2	9.6	1.5	6.6	1.2	5.4
7			2.8	10.6	1.9	7.3	1.4	5.8
8			3.4	11.7	2.3	7.9	1.7	6.2
9			4.0	12.7	2.7	8.6	2.0	6.6
10			4.6	13.7	3.1	9.2	2.3	7.1
11			5.3	14.7	3.5	9.9	2.6	7.5
12			5.9	15.6	3.9	10.5	3.0	8.0
13			6.6	16.6	4.4	11.2	3.3	8.5
14			7.2	17.6	4.8	11.8	3.6	8.9
15			7.9	18.5	5.3	12.5	3.9	9.4
16			8.6	19.5	5.7	13.1	4.3	9.9
17			9.3	20.4	6.2	13.8	4.6	10.4
18			10.0	21.4	6.6	14.4	5.0	10.8
19			10.7	22.3	7.1	15.0	5.3	11.3
20			11.4	23.2	7.5	15.7	5.6	11.8
21			12.1	24.1	8.0	16.3	6.0	12.3
22			12.8	25.1	8.5	16.9	6.3	12.7
23			13.5	26.0	9.0	17.5	6.7	13.2
24			14.3	26.9	9.4	18.1	7.1	13.7
25			15.0	27.0	9.9	18.8	7.4	14.2

*The two-step (double) sample confidence limits in this table are based on an initial sample of size n_1 with an additional sample of size n_2 *only* if occurrences are found in the initial sample.

Table A-2

Two-Step Sampling Confidence Limits*

Initial Sample (n_1) ***= 60***

Confidence Level = 90% (Two-Sided), 95% (One-Sided)

N = ***10,000***

Occurrences in Sample	Single Sample n_1		Two-Step (Double) Sample ($n_1 + n_2$)					
			n_2 = 60		n_2 = 120		n_2 = 180	
	LL	UL	LL	UL	LL	UL	LL	UL
0	0.0%	3.8%	—	—	—	—	—	—
1			0.1%	5.1%	0.1%	4.9%	0.1%	4.9%
2			0.3	5.7	0.3	4.9	0.2	4.9
3			0.7	6.6	0.5	5.2	0.4	4.9
4			1.2	7.6	0.8	5.6	0.6	5.0
5			1.7	8.6	1.1	6.1	0.9	5.2
6			2.2	9.6	1.5	6.7	1.1	5.5
7			2.8	10.7	1.9	7.3	1.4	5.8
8			3.4	11.7	2.3	7.9	1.7	6.2
9			4.0	12.7	2.7	8.6	2.0	6.6
10			4.6	13.7	3.1	9.2	2.3	7.1
11			5.2	14.7	3.5	9.9	2.6	7.6
12			5.9	15.7	3.9	10.6	2.9	8.0
13			6.5	16.7	4.3	11.2	3.3	8.5
14			7.2	17.7	4.8	11.9	3.6	9.0
15			7.9	18.6	5.2	12.5	3.9	9.5
16			8.6	19.5	5.7	13.2	4.3	9.9
17			9.3	20.5	6.1	13.8	4.6	10.4
18			10.0	21.4	6.6	14.4	4.9	10.9
19			10.6	22.3	7.0	15.1	5.3	11.4
20			11.4	23.3	7.5	15.7	5.6	11.8
21			12.1	24.2	8.0	16.3	6.0	12.3
22			12.8	25.1	8.5	16.9	6.3	12.8
23			13.5	26.0	8.9	17.6	6.7	13.3
24			14.2	26.9	9.4	18.2	7.0	13.7
25			14.9	27.8	9.9	18.8	7.4	14.2

*The two-step (double) sample confidence limits in this table are based on an initial sample of size n_1 with an additional sample of size n_2 *only* if occurrences are found in the initial sample.

Table A-2
Two-Step Sampling Confidence Limits*
Initial Sample (n_1) = 70
Confidence Level = 90% (Two -Sided), 95% (One-Sided)
N = 1000

Occurrences in Sample	Single Sample n_1		Two-Step (Double) Sample ($n_1 + n_2$)					
			$n_2 = 60$		$n_2 = 120$		$n_2 = 180$	
	LL	UL	LL	UL	LL	UL	LL	UL
0	0.0%	3.1%	—	—	—	—	—	—
1			0.1%	4.2%	0.1%	4.0%	0.1%	4.0%
2			0.3	3.9	0.2	4.1	0.2	4.0
3			0.7	5.8	0.5	4.4	0.4	4.1
4			1.2	6.7	0.8	4.8	0.7	4.2
5			1.6	7.7	1.2	5.4	0.9	4.4
6			2.2	8.6	1.5	6.0	1.2	4.7
7			2.7	9.6	1.9	6.6	1.5	5.1
8			3.3	10.5	2.3	7.2	1.8	5.5
9			3.8	11.5	2.7	7.8	2.1	6.0
10			4.4	12.4	3.1	8.4	2.4	6.4
11			5.0	13.3	3.5	9.1	2.7	6.9
12			5.6	14.2	3.9	9.7	3.0	7.3
13			6.2	15.1	4.3	10.3	3.3	7.8
14			6.9	16.0	4.7	10.9	3.7	8.2
15			7.5	16.8	5.2	11.5	4.0	8.7
16			8.1	17.7	5.6	12.1	4.3	9.2
17			8.8	18.6	6.1	12.7	4.7	9.6
18			9.4	19.4	6.5	13.3	5.0	10.1
19			10.1	20.3	6.9	13.9	5.3	10.5
20			10.7	21.2	7.4	14.5	5.7	11.0
21			11.4	22.0	7.8	15.1	6.0	11.4
22			12.0	22.9	8.3	15.7	6.4	11.9
23			12.7	23.7	8.7	16.3	6.7	12.3
24			13.4	24.5	9.2	16.8	7.0	12.8
25			14.1	25.4	9.7	17.4	7.4	13.2

*The two-step (double) sample confidence limits in this table are based on an initial sample of size n_1 with an additional sample of size n_2 *only* if occurrences are found in the initial sample.

Table A-2
*Two-Step Sampling Confidence Limits**
Initial Sample (n_1) = *70*
Confidence Level = 90% (Two-Sided), 95% (One-Sided)
N = *5000*

Occurrences in Sample	Single Sample n_1		Two-Step (Double) Sample ($n_1 + n_2$)					
			$n_2 = 60$		$n_2 = 120$		$n_2 = 180$	
	LL	UL	LL	UL	LL	UL	LL	UL
0	0.0%	3.2%	—	—	—	—	—	—
1			0.1%	4.4%	0.1%	4.2%	0.1%	4.0%
2			0.3	5.1	0.2	4.3	0.2	4.2
3			0.7	6.0	0.5	4.6	0.4	4.2
4			1.1	6.9	0.8	5.1	0.6	4.4
5			1.6	7.9	1.1	5.6	0.8	4.7
6			2.1	8.9	1.4	6.2	1.1	5.0
7			2.6	9.8	1.8	6.8	1.4	5.4
8			3.1	10.8	2.1	7.5	1.6	5.8
9			3.7	11.7	2.5	8.1	1.9	6.3
10			4.3	12.6	2.9	8.7	2.2	6.7
11			4.9	13.6	3.3	9.3	2.5	7.2
12			5.5	14.5	3.7	10.0	2.8	7.6
13			6.1	15.4	4.1	10.6	3.1	8.1
14			6.7	16.3	4.6	11.2	3.5	8.6
15			7.3	17.1	5.0	11.8	3.8	9.0
16			7.9	18.0	5.4	12.4	4.1	9.5
17			8.6	18.9	5.8	13.0	4.4	10.0
18			9.2	19.8	6.3	13.6	4.8	10.4
19			9.8	20.6	6.7	14.2	5.1	10.9
20			10.5	21.5	7.1	14.8	5.4	11.3
21			11.1	22.3	7.6	15.4	5.8	11.8
22			11.8	23.2	8.0	16.0	6.1	12.2
23			12.5	24.0	8.5	16.6	6.4	12.7
24			13.1	24.9	8.9	17.2	6.8	13.1
25			13.8	25.7	9.4	17.8	7.1	13.6

*The two-step (double) sample confidence limits in this table are based on an initial sample of size n_1 with an additional sample of size n_2 *only* if occurrences are found in the initial sample.

Table A-2
Two-Step Sampling Confidence Limits*
Initial Sample (n_1) ***= 70***
Confidence Level = 90% (Two-Sided), 95% (One-Sided)
N = ***10,000***

Occurrences in Sample	Single Sample n_1		Two-Step (Double) Sample ($n_1 + n_2$)					
			$n_2 = 60$		$n_2 = 120$		$n_2 = 180$	
	LL	UL	LL	UL	LL	UL	LL	UL
0	0.0%	3.2%	—	—	—	—	—	—
1			0.1%	4.4%	0.1%	4.2%	0.1%	4.2%
2			0.3	5.1	0.2	4.3	0.2	4.2
3			0.7	6.0	0.5	4.6	0.4	4.3
4			1.1	6.9	0.8	5.1	0.6	4.4
5			1.5	7.9	1.1	5.6	0.8	4.7
6			2.0	8.9	1.4	6.2	1.1	5.0
7			2.6	9.9	1.8	6.9	1.4	5.4
8			3.1	10.8	2.1	7.5	1.6	5.8
9			3.7	11.7	2.5	8.1	1.9	6.3
10			4.2	12.7	2.9	8.7	2.2	6.7
11			4.8	13.6	3.3	9.4	2.5	7.2
12			5.4	14.5	3.7	10.0	2.8	7.7
13			6.0	15.4	4.1	10.6	3.1	8.1
14			6.7	16.3	4.5	11.2	3.4	8.6
15			7.3	17.2	5.0	11.9	3.8	9.1
16			7.9	18.1	5.4	12.5	4.1	9.5
17			8.5	18.9	5.8	13.1	4.4	10.0
18			9.2	19.8	6.2	13.7	4.7	10.5
19			9.8	20.7	6.7	14.3	5.1	10.9
20			10.5	21.5	7.1	14.9	5.4	11.4
21			11.1	22.4	7.6	15.5	5.7	11.8
22			11.8	23.2	8.0	16.1	6.1	12.3
23			12.4	24.1	8.4	16.7	6.4	12.7
24			13.1	24.9	8.9	17.3	6.7	13.2
25			13.8	25.8	9.3	17.8	7.1	13.6

*The two-step (double) sample confidence limits in this table are based on an initial sample of size n_1 with an additional sample of size n_2 *only* if occurrences are found in the initial sample.

Table A-2
Two-Step Sampling Confidence Limits*
Initial Sample (n_1) ***= 80***
Confidence Level = 90% (Two-Sided), 95% (One-Sided)
N = ***1000***

Occurrences in Sample	Single Sample n_1		Two-Step (Double) Sample $(n_1 + n_2)$					
			$n_2 = 60$		$n_2 = 120$		$n_2 = 180$	
	LL	UL	LL	UL	LL	UL	LL	UL
0	0.0%	2.7%	—	—	—	—	—	—
1			0.1%	3.8%	0.1%	3.5%	0.1%	3.5%
2			0.3	4.4	0.2	3.7	0.2	3.5
3			0.7	5.3	0.5	4.0	0.4	3.6
4			1.1	6.2	0.8	4.5	0.6	3.8
5			1.5	7.1	1.1	5.0	0.9	4.1
6			2.0	8.0	1.5	5.6	1.2	4.4
7			2.5	8.9	1.8	6.2	1.4	4.8
8			3.0	9.8	2.2	6.8	1.7	5.2
9			3.6	10.6	2.5	7.4	2.0	5.7
10			4.1	11.5	2.9	8.0	2.3	6.1
11			4.7	12.3	3.3	8.6	2.6	6.6
12			5.2	13.2	3.7	9.2	2.9	7.0
13			5.8	14.0	4.1	9.8	3.2	7.5
14			6.4	14.8	4.5	10.3	3.5	7.9
15			7.0	15.6	4.9	10.9	3.8	8.3
16			7.6	16.5	5.3	11.5	4.2	8.8
17			8.2	17.3	5.8	12.1	4.5	9.2
18			8.8	18.1	6.2	12.6	4.8	9.7
19			9.4	18.9	6.6	13.2	5.1	10.1
20			10.0	19.7	7.0	13.8	5.5	10.5
21			10.6	20.5	7.5	14.3	5.8	11.0
22			11.2	21.3	7.9	14.9	6.1	11.4
23			11.8	22.0	8.3	15.4	6.5	11.8
24			12.4	22.8	8.7	16.0	6.8	12.3
25			13.1	23.6	9.2	16.6	7.1	12.7

*The two-step (double) sample confidence limits in this table are based on an initial sample of size n_1 with an additional sample of size n_2 *only* if occurrences are found in the initial sample.

Table A-2
Two-Step Sampling Confidence Limits*
Initial Sample (n_1) **= *80***
Confidence Level = 90% (Two-Sided), 95% (One-Sided)
N = ***5000***

Occurrences in Sample	Single Sample n_1		Two-Step (Double) Sample $(n_1 + n_2)$					
			n_2 = 60		n_2 = 120		n_2 = 180	
	LL	UL	LL	UL	LL	UL	LL	UL
0	0.0%	2.8%	—	—	—	—	—	—
1			0.1%	3.9%	0.1%	3.7%	0.1%	3.7%
2			0.3	4.6	0.2	3.9	0.2	3.7
3			0.6	5.5	0.5	4.2	0.4	3.8
4			1.0	6.4	0.7	4.7	0.6	4.0
5			1.4	7.3	1.0	5.3	0.8	4.3
6			1.9	8.2	1.3	5.9	1.1	4.7
7			2.4	9.1	1.7	6.5	1.3	5.1
8			2.9	10.0	2.0	7.1	1.6	5.5
9			3.4	10.9	2.4	7.7	1.9	6.0
10			4.0	11.8	2.8	8.3	2.1	6.4
11			4.5	12.6	3.2	8.9	2.4	6.9
12			5.1	13.5	3.5	9.5	2.7	7.3
13			5.6	14.3	3.9	10.1	3.0	7.8
14			6.2	15.1	4.3	10.7	3.3	8.2
15			6.8	15.9	4.7	11.2	3.6	8.7
16			7.3	16.8	5.1	11.8	3.9	9.1
17			7.9	17.6	5.5	12.4	4.3	9.6
18			8.5	18.4	6.0	13.0	4.6	10.0
19			9.1	19.2	6.4	13.5	4.9	10.5
20			9.7	20.0	6.8	14.1	5.2	10.9
21			10.3	20.8	7.2	14.7	5.5	11.3
22			10.9	21.6	7.6	15.2	5.9	11.8
23			11.6	22.4	8.0	15.8	6.2	12.2
24			12.2	23.2	8.5	16.4	6.5	12.6
25			12.8	24.0	8.9	16.9	6.8	13.1

*The two-step (double) sample confidence limits in this table are based on an initial sample of size n_1 with an additional sample of size n_2 *only* if occurrences are found in the initial sample.

Table A-2

Two-Step Sampling Confidence Limits*

Initial Sample* (n_1) = *80

Confidence Level = 90% (Two-Sided), 95% (One-Sided)

N = ***10,000***

Occurrences in Sample	Single Sample n_1		Two-Step (Double) Sample ($n_1 + n_2$)					
			$n_2 = 60$		$n_2 = 120$		$n_2 = 180$	
	LL	UL	LL	UL	LL	UL	LL	UL
0	0.0%	2.8%	—	—	—	—	—	—
1			0.1%	4.0%	0.1%	3.7%	0.1%	3.7%
2			0.3	4.7	0.2	3.9	0.2	3.7
3			0.6	5.5	0.5	4.2	0.4	3.8
4			1.0	6.4	0.7	4.7	0.6	4.0
5			1.4	7.4	1.0	5.3	0.8	4.3
6			1.9	8.3	1.3	5.9	1.0	4.7
7			2.4	9.2	1.7	6.5	1.3	5.1
8			2.9	10.0	2.0	7.1	1.6	5.6
9			3.4	10.9	2.4	7.7	1.8	6.0
10			3.9	11.8	2.8	8.3	2.1	6.5
11			4.5	12.6	3.1	8.9	2.4	6.9
12			5.0	13.5	3.5	9.5	2.7	7.4
13			5.6	14.3	3.9	10.1	3.0	7.8
14			6.2	15.2	4.3	10.7	3.3	8.3
15			6.7	16.0	4.7	11.3	3.6	8.7
16			7.3	16.8	5.1	11.9	3.9	9.2
17			7.9	17.6	5.5	12.4	4.2	9.6
18			8.5	18.4	5.9	13.0	4.5	10.1
19			9.1	19.2	6.3	13.6	4.9	10.5
20			9.7	20.0	6.8	14.2	5.2	10.9
21			10.3	20.8	7.2	14.7	5.5	11.4
22			10.9	21.6	7.6	15.3	5.8	11.8
23			11.5	22.4	8.0	15.9	6.2	12.3
24			12.1	23.2	8.4	16.4	6.5	12.7
25			12.8	24.0	8.9	17.0	6.8	13.1

*The two-step (double) sample confidence limits in this table are based on an initial sample of size n_1 with an additional sample of size n_2 *only* if occurrences are found in the initial sample.

Table A-2
Two-Step Sampling Confidence Limits*
Initial Sample (n_1) = 90
Confidence Level = 90% (Two-Sided), 95% (One-Sided)
N = _1000_

Occurrences in Sample	Single Sample n_1		Two-Step (Double) Sample (n_1 + n_2)					
			n_2 = 60		n_2 = 120		n_2 = 180	
	LL	UL	LL	UL	LL	UL	LL	UL
0	0.0%	2.4%	—	—	—	—	—	—
1			0.1%	3.4%	0.1%	3.1%	0.1%	3.1%
2			0.2	4.1	0.2	3.3	0.2	3.1
3			0.6	4.9	0.5	3.7	0.4	3.2
4			1.0	5.8	0.8	4.2	0.6	3.5
5			1.4	6.6	1.1	4.7	0.9	3.8
6			1.9	7.5	1.4	5.3	1.1	4.2
7			2.4	8.3	1.7	5.9	1.4	4.6
8			2.8	9.1	2.1	6.5	1.7	5.0
9			3.3	9.9	2.4	7.0	1.9	5.4
10			3.8	10.7	2.8	7.6	2.2	5.9
11			4.4	11.5	3.2	8.2	2.5	6.3
12			4.9	12.3	3.5	8.7	2.8	6.7
13			5.4	13.1	3.9	9.3	3.1	7.2
14			6.0	13.8	4.3	9.8	3.4	7.6
15			6.5	14.6	4.7	10.4	3.7	8.0
16			7.1	15.4	5.1	10.9	4.0	8.4
17			7.6	16.1	5.5	11.5	4.3	8.9
18			8.2	16.9	5.9	12.0	4.6	9.3
19			8.7	17.6	6.3	12.6	5.0	9.7
20			9.3	18.4	6.7	13.1	5.3	10.1
21			9.9	19.1	7.1	13.6	5.6	10.5
22			10.5	19.9	7.5	14.2	5.9	11.0
23			11.0	20.6	7.9	14.7	6.2	11.4
24			11.6	21.3	8.3	15.2	6.5	11.8
25			12.2	22.1	8.8	15.8	6.9	12.2

*The two-step (double) sample confidence limits in this table are based on an initial sample of size n_1 with an additional sample of size n_2 *only* if occurrences are found in the initial sample.

Table A-2
Two-Step Sampling Confidence Limits*
Initial Sample (n_1) = 90
Confidence Level = 90% (Two-Sided), 95% (One-Sided)
N = **_5000_**

Occurrences in Sample	Single Sample n_1		Two-Step (Double) Sample ($n_1 + n_2$)					
			n_2 = 60		n_2 = 120		n_2 = 180	
	LL	UL	LL	UL	LL	UL	LL	UL
0	0.0%	2.5%	—	—	—	—	—	—
1			0.1%	3.6%	0.1%	3.3%	0.1%	3.3%
2			0.3	4.3	0.2	3.5	0.2	3.3
3			0.6	5.1	0.4	3.9	0.4	3.4
4			0.9	6.0	0.7	4.4	0.6	3.7
5			1.3	6.8	1.0	5.0	0.9	4.1
6			1.8	7.7	1.3	5.6	1.0	4.4
7			2.2	8.5	1.6	6.1	1.3	4.9
8			2.7	9.4	1.9	6.7	1.5	5.3
9			3.2	10.2	2.3	7.3	1.8	5.7
10			3.7	11.0	2.6	7.9	2.1	6.2
11			4.2	11.8	3.0	8.5	2.3	6.6
12			4.7	12.6	3.4	9.0	2.6	7.0
13			5.2	13.4	3.7	9.6	2.9	7.5
14			5.8	14.1	4.1	10.2	3.2	7.9
15			6.3	14.9	4.5	10.7	3.5	8.4
16			6.9	15.7	4.9	11.3	3.8	8.8
17			7.4	16.4	5.3	11.8	4.1	9.2
18			8.0	17.2	5.7	12.4	4.4	9.6
19			8.5	18.0	6.1	12.9	4.7	10.1
20			9.1	18.7	6.5	13.5	5.0	10.5
21			9.6	19.5	6.9	14.0	5.3	10.9
22			10.2	20.2	7.3	14.5	5.6	11.3
23			10.8	20.9	7.7	15.1	6.0	11.8
24			11.3	21.7	8.1	15.6	6.3	12.2
25			11.9	22.4	8.5	16.1	6.6	12.6

*The two-step (double) sample confidence limits in this table are based on an initial sample of size n_1 with an additional sample of size n_2 *only* if occurrences are found in the initial sample.

Table A-2

Two-Step Sampling Confidence Limits*

Initial Sample (n_1) **= *90***

Confidence Level = 90% (Two-Sided), 95% (One-Sided)

N = ***10,000***

Occurrences in Sample	Single Sample n_1		Two-Step (Double) Sample ($n_1 + n_2$)					
			$n_2 = 60$		$n_2 = 120$		$n_2 = 180$	
	LL	UL	LL	UL	LL	UL	LL	UL
0	0.0%	2.5%	—	—	—	—	—	—
1			0.1%	3.6%	0.1%	3.3%	0.1%	3.3%
2			0.3	4.3	0.2	3.5	0.2	3.3
3			0.6	5.1	0.4	3.9	0.3	3.5
4			0.9	6.0	0.7	4.4	0.5	3.7
5			1.3	6.9	1.0	5.0	0.8	4.1
6			1.8	7.7	1.3	5.6	1.0	4.5
7			2.2	8.6	1.6	6.2	1.2	4.9
8			2.7	9.4	1.9	6.8	1.5	5.3
9			3.2	10.2	2.3	7.3	1.8	5.8
10			3.7	11.0	2.6	7.9	2.0	6.2
11			4.2	11.8	3.0	8.5	2.3	6.6
12			4.7	12.6	3.3	9.1	2.6	7.1
13			5.2	13.4	3.7	9.6	2.9	7.5
14			5.8	14.2	4.1	10.2	3.2	8.0
15			6.3	14.9	4.5	10.7	3.5	8.4
16			6.8	15.7	4.9	11.3	3.8	8.8
17			7.4	16.5	5.2	11.9	4.1	9.3
18			7.9	17.2	5.6	12.4	4.4	9.7
19			8.5	18.0	6.0	13.0	4.7	10.1
20			9.0	18.7	6.4	13.5	5.0	10.5
21			9.6	19.5	6.8	14.0	5.3	11.0
22			10.2	20.2	7.2	14.6	5.6	11.4
23			10.7	21.0	7.6	15.1	5.9	11.8
24			11.3	21.7	8.0	15.7	6.2	12.2
25			11.9	22.5	8.4	16.2	6.6	12.6

*The two-step (double) sample confidence limits in this table are based on an initial sample of size n_1 with an additional sample of size n_2 *only* if occurences are found in the initial sample.

Table A-2
Two-Step Sampling Confidence Limits*
Initial Sample** (n_1) **= 100
Confidence Level = 90% (Two-Sided), 95% (One-Sided)
N = ***1000***

Occurrences in Sample	Single Sample n_1		Two-Step (Double) Sample ($n_1 + n_2$)					
			$n_2 = 60$		$n_2 = 120$		$n_2 = 180$	
	LL	UL	LL	UL	LL	UL	LL	UL
0	0.0%	2.2%	—	—	—	—	—	—
1			0.1%	3.1%	0.1%	2.8%	0.1%	2.8%
2			0.2	3.8	0.2	3.0	0.2	2.8
3			0.6	4.6	0.5	3.5	0.3	3.0
4			0.9	5.4	0.7	4.0	0.6	3.2
5			1.3	6.2	1.0	4.5	0.8	3.6
6			1.8	7.0	1.3	5.0	1.1	4.0
7			2.2	7.8	1.7	5.6	1.3	4.4
8			2.7	8.5	2.0	6.1	1.6	4.8
9			3.1	9.3	2.3	6.7	1.9	5.2
10			3.6	10.0	2.7	7.2	2.1	5.6
11			4.1	10.8	3.0	7.8	2.4	6.1
12			4.6	11.5	3.4	8.3	2.7	6.5
13			5.1	12.2	3.8	8.9	3.0	6.9
14			5.6	13.0	4.1	9.4	3.3	7.3
15			6.1	13.7	4.5	9.9	3.6	7.7
16			6.6	14.4	4.9	10.4	3.9	8.1
17			7.1	15.1	5.3	11.0	4.2	8.5
18			7.7	15.8	5.6	11.5	4.5	8.9
19			8.2	16.5	6.0	12.0	4.8	9.4
20			8.7	17.2	6.4	12.5	5.1	9.8
21			9.3	17.9	6.8	13.0	5.4	10.2
22			9.8	18.6	7.2	13.5	5.7	10.6
23			10.3	19.3	7.6	14.0	6.0	11.0
24			10.9	20.0	8.0	14.5	6.3	11.4
25			11.4	20.7	8.4	15.0	6.6	11.8

*The two-step (double) sample confidence limits in this table are based on an initial sample of size n_1 with an additional sample of size n_2 *only* if occurrences are found in the initial sample.

Table A-2

Two-Step Sampling Confidence Limits*

Initial Sample* (n_1) = *100

Confidence Level = 90% (Two-Sided), 95% (One-Sided)

N = ***5000***

Occurrences in Sample	Single Sample n_1		Two-Step (Double) Sample ($n_1 + n_2$)					
			$n_2 = 60$		$n_2 = 120$		$n_2 = 180$	
	LL	UL	LL	UL	LL	UL	LL	UL
0	0.0%	2.2%	—	—	—	—	—	—
1			0.1%	3.3%	0.1%	3.0%	0.1%	2.9%
2			0.2	4.0	0.2	3.2	0.2	3.0
3			0.5	4.8	0.4	3.7	0.3	3.2
4			0.9	5.6	0.7	4.2	0.5	3.5
5			1.3	6.4	0.9	4.7	0.7	3.8
6			1.7	7.2	1.2	5.3	1.0	4.2
7			2.1	8.0	1.5	5.8	1.2	4.7
8			2.5	8.8	1.9	6.4	1.5	5.1
9			3.0	9.5	2.2	7.0	1.7	5.5
10			3.5	10.3	2.5	7.5	2.0	5.9
11			3.9	11.1	2.9	8.1	2.3	6.4
12			4.4	11.8	3.2	8.6	2.5	6.8
13			4.9	12.5	3.6	9.2	2.8	7.2
14			5.4	13.3	3.9	9.7	3.1	7.6
15			5.9	14.0	4.3	10.2	3.4	8.1
16			6.4	14.7	4.7	10.8	3.7	8.5
17			6.9	15.4	5.0	11.3	4.0	8.9
18			7.5	16.1	5.4	11.8	4.2	9.3
19			8.0	16.9	5.8	12.3	4.5	9.7
20			8.5	17.6	6.2	12.9	4.8	10.1
21			9.0	18.3	6.5	13.4	5.1	10.5
22			9.6	19.0	6.9	13.9	5.4	10.9
23			10.1	19.7	7.3	14.4	5.7	11.3
24			10.6	20.4	7.7	14.9	6.0	11.8
25			11.2	21.0	8.1	15.4	6.3	12.2

*The two-step (double) sample confidence limits in this table are based on an initial sample of size n_1 with an additional sample of size n_2 *only* if occurrences are found in the initial sample.

Table A-2
Two-Step Sampling Confidence Limits*
Initial Sample (n_1) ***= 100***
Confidence Level = 90% (Two-Sided), 95% (One-Sided)
N = ***10,000***

Occurrences in Sample	Single Sample n_1		Two-Step (Double) Sample ($n_1 + n_2$)					
			$n_2 = 60$		$n_2 = 120$		$n_2 = 180$	
	LL	UL	LL	UL	LL	UL	LL	UL
0	0.0%	2.3%	—	—	—	—	—	—
1			0.1%	3.3%	0.1%	3.0%	0.1%	2.9%
2			0.2	4.0	0.2	3.3	0.2	3.0
3			0.5	4.8	0.4	3.7	0.3	3.2
4			0.9	5.6	0.6	4.2	0.5	3.5
5			1.2	6.4	0.9	4.8	0.7	3.9
6			1.7	7.2	1.2	5.3	1.0	4.3
7			2.1	8.0	1.5	5.9	1.2	4.7
8			2.5	8.8	1.8	6.4	1.4	5.1
9			3.0	9.6	2.2	7.0	1.7	5.5
10			3.4	10.3	2.5	7.6	2.0	6.0
11			3.9	11.1	2.8	8.1	2.2	6.4
12			4.4	11.8	3.2	8.7	2.5	6.8
13			4.9	12.6	3.6	9.2	2.8	7.2
14			5.4	13.3	3.9	9.7	3.1	7.7
15			5.9	14.0	4.3	10.3	3.4	8.1
16			6.4	14.8	4.6	10.8	3.6	8.5
17			6.9	15.5	5.0	11.3	3.9	8.9
18			7.4	16.2	5.4	11.9	4.2	9.3
19			7.9	16.9	5.8	12.4	4.5	9.8
20			8.5	17.6	6.1	12.9	4.8	10.2
21			9.0	18.3	6.5	13.4	5.1	10.6
22			9.5	19.0	6.9	13.9	5.4	11.0
23			10.1	19.7	7.3	14.4	5.7	11.4
24			10.6	20.4	7.7	15.0	6.0	11.8
25			11.1	21.1	8.1	15.5	6.3	12.2

*The two-step (double) sample confidence limits in this table are based on an initial sample of size n_1 with an additional sample of size n_2 *only* if occurrences are found in the initial sample.

Table A-2
Two-Step Sampling Confidence Limits*
Initial Sample (n_1) ***= 50***
Confidence Level = 95% (Two-Sided), 97.5% (One-Sided)
N = ***1000***

Occurrences in Sample	Single Sample n_1		Two-Step (Double) Sample $(n_1 + n_2)$					
			$n_2 = 60$		$n_2 = 120$		$n_2 = 180$	
	LL	UL	LL	UL	LL	UL	LL	UL
0	0.0%	5.7%	—	—	—	—	—	—
1			0.1%	7.0%	0.1%	6.9%	0.1%	6.9%
2			0.3	7.3	0.2	6.9	0.2	6.9
3			0.7	8.0	0.5	7.0	0.4	6.9
4			1.1	9.0	0.8	7.1	0.6	6.9
5			1.6	10.1	1.1	7.4	0.9	6.9
6			2.2	11.2	1.5	7.8	1.2	7.0
7			2.8	12.4	1.9	8.3	1.4	7.1
8			3.4	13.5	2.3	8.9	1.7	7.3
9			4.0	14.6	2.7	9.6	2.0	7.6
10			4.7	15.7	3.1	10.3	2.3	7.9
11			5.3	16.8	3.5	10.9	2.7	8.3
12			6.0	17.8	4.0	11.6	3.0	8.8
13			6.7	18.9	4.4	12.3	3.3	9.2
14			7.4	20.0	4.9	13.0	3.7	9.7
15			8.1	21.1	5.3	13.7	4.0	10.2
16			8.8	22.1	5.8	14.4	4.3	10.6
17			9.6	23.1	6.3	15.1	4.7	11.1
18			10.3	24.2	6.7	15.7	5.0	11.6
19			11.1	25.2	7.2	16.4	5.4	12.1
20			11.8	26.2	7.7	17.1	5.8	12.6
21			12.6	27.2	8.2	17.7	6.1	13.1
22			13.3	28.2	8.7	18.4	6.5	13.6
23			14.1	29.2	9.2	19.1	6.8	14.1
24			14.9	30.2	9.6	19.7	7.2	14.6
25			15.7	31.2	10.1	20.4	7.6	15.1

*The two-step (double) sample confidence limits in this table are based on an initial sample of size n_1 with an additional sample of size n_2 *only* if occurrences are found in the initial sample.

Table A-2
Two-Step Sampling Confidence Limits*
Initial Sample (n_1) ***= 50***
Confidence Level = 95% (Two-Sided), 97.5% (One-Sided)
N = ***5000***

Occurrences in Sample	Single Sample n_1		Two-Step (Double) Sample ($n_1 + n_2$)					
			n_2 = 60		n_2 = 120		n_2 = 180	
	LL	UL	LL	UL	LL	UL	LL	UL
0	0.0%	5.8%	—	—	—	—	—	—
1			0.1%	7.2%	0.1%	7.1%	0.1%	7.1%
2			0.3	7.5	0.2	7.1	0.2	7.1
3			0.6	8.3	0.4	7.1	0.3	7.1
4			1.0	9.3	0.7	7.3	0.6	7.1
5			1.5	10.4	1.0	7.6	0.8	7.1
6			2.1	11.5	1.4	8.1	1.0	7.2
7			2.6	12.6	1.7	8.6	1.3	7.3
8			3.2	13.8	2.1	9.3	1.6	7.4
9			3.8	14.9	2.5	9.9	1.9	7.9
10			4.5	16.0	2.9	10.6	2.2	8.3
11			5.1	17.1	3.3	11.3	2.5	8.7
12			5.8	18.8	3.8	12.0	2.8	9.1
13			6.5	19.2	4.2	12.7	3.1	9.6
14			7.2	20.3	4.6	13.4	3.4	10.1
15			7.9	21.4	5.1	14.1	3.8	10.6
16			8.6	22.5	5.5	14.8	4.1	11.0
17			9.3	23.5	6.0	15.4	4.4	11.5
18			10.1	24.5	6.8	16.1	4.8	12.0
19			10.8	25.6	6.9	16.8	5.1	12.5
20			11.5	26.6	7.4	17.5	5.5	13.0
21			12.3	27.6	7.9	18.2	5.8	13.5
22			13.0	28.6	8.4	18.8	6.2	14.2
23			13.8	29.6	8.8	19.5	6.5	14.5
24			14.6	30.6	9.3	20.2	6.9	15.0
25			15.4	31.6	9.8	20.8	7.2	15.5

*The two-step (double) sample confidence limits in this table are based on an initial sample of size n_1 with an additional sample of size n_2 *only* if occurrences are found in the initial sample.

Table A-2
Two-Step Sampling Confidence Limits*
Initial Sample (n_1) ***= 50***
Confidence Level = 95% (Two-Sided), 97.5% (One-Sided)
N = ***10,000***

Occurrences in Sample	Single Sample n_1		Two-Step (Double) Sample ($n_1 + n_2$)					
			$n_2 = 60$		$n_2 = 120$		$n_2 = 180$	
	LL	UL	LL	UL	LL	UL	LL	UL
0	0.0%	5.8%	—	—	—	—	—	—
1			0.1%	7.2%	0.1%	7.2%	0.1%	7.1%
2			0.3	7.6	0.2	7.1	0.2	7.1
3			0.6	8.3	0.4	7.2	0.3	7.1
4			1.0	9.3	0.7	7.3	0.5	7.1
5			1.5	10.4	1.0	7.7	0.8	7.1
6			2.0	11.5	1.4	8.1	1.0	7.2
7			2.6	12.7	1.7	8.7	1.3	7.4
8			3.2	13.8	2.1	9.3	1.6	7.6
9			3.8	14.9	2.5	10.0	1.9	7.9
10			4.5	16.0	2.9	10.6	2.1	8.3
11			5.1	17.2	3.3	11.3	2.5	8.7
12			5.8	18.1	3.7	12.0	2.8	9.2
13			6.5	19.2	4.2	12.7	3.1	9.6
14			7.2	20.3	4.6	13.4	3.4	10.1
15			7.9	21.5	5.1	14.1	3.7	10.6
16			8.8	22.5	5.5	14.8	4.1	11.1
17			9.3	23.5	6.0	15.5	4.4	11.6
18			10.0	24.6	6.4	16.2	4.7	12.1
19			10.8	25.6	6.9	16.9	5.1	12.6
20			11.5	26.6	7.4	17.5	5.4	13.1
21			12.3	27.6	7.8	18.2	5.8	13.6
22			13.0	28.7	8.3	18.9	6.1	14.1
23			13.8	29.7	8.8	19.6	6.5	14.6
24			14.5	30.7	9.3	20.2	6.8	15.1
25			15.3	31.7	9.8	20.9	7.2	15.6

*The two-step (double) sample confidence limits in this table are based on an initial sample of size n_1 with an additional sample of size n_2 *only* if occurrences are found in the initial sample.

Table A-2

Two-Step Sampling Confidence Limits*

Initial Sample (n_1) ***= 60***

Confidence Level = 95% (Two-Sided), 97.5% (One-Sided)

N = ***1000***

Occurrences in Sample	Single Sample n_1		Two-Step (Double) Sample ($n_1 + n_2$)					
			$n_2 = 60$		$n_2 = 120$		$n_2 = 180$	
	LL	UL	LL	UL	LL	UL	LL	UL
0	0.0%	4.7%	—	—	—	—	—	—
1			0.1%	5.9%	0.1%	5.8%	0.1%	5.8%
2			0.2	6.3	0.2	5.8	0.2	5.8
3			0.6	7.2	0.5	5.9	0.3	5.8
4			1.0	8.1	0.7	6.1	0.6	5.8
5			1.5	9.2	1.1	6.6	0.8	5.8
6			2.0	10.3	1.4	7.1	1.1	6.0
7			2.5	11.3	1.8	7.7	1.4	6.2
8			3.1	12.4	2.1	8.3	1.7	6.5
9			3.7	13.4	2.5	9.0	2.0	6.9
10			4.3	14.4	2.9	9.6	2.3	7.3
11			4.9	15.4	3.3	10.3	2.6	7.8
12			5.5	16.4	3.7	10.9	2.9	8.2
13			6.2	17.3	4.2	11.6	3.2	8.7
14			6.8	18.3	4.6	12.3	3.5	9.2
15			7.5	19.3	5.0	12.9	3.8	9.7
16			8.1	20.3	5.5	13.6	4.2	10.1
17			8.8	21.3	5.9	14.2	4.5	10.6
18			9.5	22.2	6.4	14.9	4.8	11.1
19			10.1	23.1	6.8	15.5	5.2	11.6
20			10.8	24.1	7.3	16.1	5.5	12.1
21			11.5	25.0	7.7	16.8	5.9	12.5
22			12.2	25.9	8.2	17.4	6.2	13.0
23			12.9	26.9	8.7	18.0	6.6	13.5
24			13.6	27.8	9.1	18.6	6.9	13.9
25			14.3	28.7	9.6	19.3	7.3	14.4

*The two-step (double) sample confidence limits in this table are based on an initial sample of size n_1 with an additional sample of size n_2 *only* if occurrences are found in the initial sample.

Table A-2

Two-Step Sampling Confidence Limits*

Initial Sample (n_1) ***= 60***

Confidence Level = 95% (Two-Sided), 97.5% (One-Sided)

N = ***5000***

Occurrences in Sample	Single Sample n_1		Two-Step (Double) Sample ($n_1 + n_2$)					
			$n_2 = 60$		$n_2 = 120$		$n_2 = 180$	
	LL	UL	LL	UL	LL	UL	LL	UL
0	0.0%	4.8%	—	—	—	—	—	—
1			0.1%	6.1%	0.1%	5.9%	0.1%	5.9%
2			0.2	6.6	0.2	6.0	0.2	5.9
3			0.6	7.4	0.4	6.1	0.3	5.9
4			0.9	8.4	0.7	6.4	0.5	6.0
5			1.4	9.5	1.0	6.8	0.7	6.1
6			1.9	10.5	1.3	7.4	1.0	6.2
7			2.4	11.6	1.6	8.0	1.2	6.5
8			3.0	12.6	2.0	8.6	1.5	6.9
9			3.5	13.7	2.4	9.3	1.8	7.2
10			4.1	14.7	2.7	9.9	2.1	7.7
11			4.7	15.7	3.1	10.6	2.4	8.1
12			5.3	16.7	3.5	11.3	2.7	8.6
13			5.9	17.7	4.0	12.0	3.0	9.1
14			6.6	18.7	4.4	12.6	3.3	9.6
15			7.2	19.7	4.8	13.3	3.6	10.1
16			7.9	20.7	5.2	13.9	3.9	10.5
17			8.5	21.6	5.7	14.6	4.2	11.5
18			9.2	22.6	6.1	15.3	4.9	12.2
19			9.9	23.5	6.5	15.9	5.2	12.5
20			10.6	24.5	7.0	16.5	5.6	13.2
21			11.2	25.4	7.4	17.2	5.9	13.4
22			11.9	26.3	7.9	17.8	6.2	13.9
23			12.6	27.3	8.3	18.5	6.6	14.4
24			13.3	28.2	8.8	19.1	6.9	14.9
25			14.0	29.1	9.3	19.7	7.3	15.4

*The two-step (double) sample confidence limits in this table are based on an initial sample of size n_1 with an additional sample of size n_2 *only* if occurrences are found in the initial sample.

Table A-2

Two-Step Sampling Confidence Limits*

Initial Sample (n_1) = ***60***

Confidence Level = 95% (Two-Sided), 97.5% (One-Sided)

N = ***10,000***

Occurrences in Sample	Single Sample n_1		Two-Step (Double) Sample ($n_1 + n_2$)					
			$n_2 = 60$		$n_2 = 120$		$n_2 = 180$	
	LL	UL	LL	UL	LL	UL	LL	UL
0	0.0%	4.9%	—	—	—	—	—	—
1			0.1%	6.1%	0.1%	5.9%	0.1%	5.9%
2			0.2	6.6	0.2	6.0	0.2	5.9
3			0.5	7.4	0.4	6.1	0.3	6.0
4			0.9	8.4	0.7	6.4	0.5	6.0
5			1.4	9.5	0.9	6.9	0.7	6.1
6			1.9	10.6	1.3	7.4	1.0	6.3
7			2.4	11.6	1.6	8.0	1.2	6.5
8			2.9	12.7	2.0	8.7	1.5	6.9
9			3.5	13.7	2.3	9.3	1.8	7.3
10			4.1	14.8	2.7	10.0	2.0	7.7
11			4.7	15.8	3.1	10.7	2.3	8.2
12			5.3	16.8	3.5	11.3	2.6	8.6
13			5.9	17.8	3.9	12.0	2.9	9.1
14			6.6	18.8	4.3	12.7	3.3	9.6
15			7.2	19.7	4.8	13.3	3.6	10.1
16			7.8	20.7	5.2	14.0	3.9	10.6
17			8.5	21.7	5.6	14.6	4.2	11.1
18			9.2	22.6	6.1	15.3	4.5	11.6
19			9.8	23.6	6.5	15.9	4.9	12.0
20			10.5	24.5	7.0	16.6	5.2	12.5
21			11.2	25.5	7.4	17.2	5.5	13.0
22			11.9	26.4	7.9	17.9	5.9	13.5
23			12.6	27.2	8.3	18.5	6.2	14.0
24			13.3	28.2	8.8	19.1	6.5	14.5
25			14.0	29.2	9.2	19.8	6.9	14.9

*The two-step (double) sample confidence limits in this table are based on an initial sample of size n_1 with an additional sample of size n_2 *only* if occurrences are found in the initial sample.

Table A-2
Two-Step Sampling Confidence Limits*
Initial Sample (n_1) ***= 70***
Confidence Level = 95% (Two-Sided), 97.5% (One-Sided)
N = ***1000***

Occurrences in Sample	Single Sample n_1		Two-Step (Double) Sample $(n_1 + n_2)$					
			$n_2 = 60$		$n_2 = 120$		$n_2 = 180$	
	LL	UL	LL	UL	LL	UL	LL	UL
0	0.0%	4.0%	—	—	—	—	—	—
1			0.1%	5.1%	0.1%	5.0%	0.1%	5.0%
2			0.2	5.6	0.2	5.0	0.2	5.0
3			0.6	6.5	0.4	5.4	0.3	5.0
4			1.0	7.5	0.7	5.5	0.6	5.0
5			1.4	8.5	1.0	6.0	0.8	5.1
6			1.9	9.4	1.3	6.6	1.1	5.4
7			2.4	10.4	1.7	7.2	1.3	5.7
8			2.9	11.4	2.0	7.8	1.6	6.1
9			3.4	12.4	2.4	8.4	1.9	6.5
10			4.0	13.3	2.8	9.1	2.2	6.9
11			4.5	14.2	3.2	9.7	2.5	7.4
12			5.1	15.1	3.6	10.3	2.8	7.8
13			5.7	16.0	4.0	11.0	3.1	8.3
14			6.3	16.9	4.4	11.6	3.4	8.8
15			6.9	17.9	4.8	12.2	3.7	9.2
16			7.5	18.4	5.2	12.8	4.0	9.7
17			8.1	19.6	5.6	13.5	4.3	10.2
18			8.7	20.5	6.0	14.1	4.7	10.6
19			9.4	21.4	6.5	14.7	5.0	11.1
20			10.0	22.3	6.9	15.3	5.3	11.6
21			10.6	23.1	7.3	15.9	5.6	12.0
22			11.3	24.0	7.8	16.5	6.0	12.5
23			11.9	24.9	8.2	17.1	6.3	12.9
24			12.6	25.7	8.7	17.7	6.7	13.4
25			13.2	26.6	9.1	18.3	7.0	13.8

*The two-step (double) sample confidence limits in this table are based on an initial sample of size n_1 with an additional sample of size n_2 *only* if occurrences are found in the initial sample.

Table A-2
Two-Step Sampling Confidence Limits*
Initial Sample (n_1) = 70
Confidence Level = 95% (Two-Sided), 97.5% (One-Sided)
N = **5000**

Occurrences in Sample	Single Sample n_1		Two-Step (Double) Sample ($n_1 + n_2$)					
			$n_2 = 60$		$n_2 = 120$		$n_2 = 180$	
	LL	UL	LL	UL	LL	UL	LL	UL
0	0.0%	4.1%	—	—	—	—	—	—
1			0.1%	5.3%	0.1%	5.1%	0.1%	5.1%
2			0.2	5.9	0.2	5.2	0.1	5.1
3			0.5	6.7	0.4	5.4	0.3	5.3
4			0.9	7.7	0.6	5.8	0.5	5.3
5			1.3	8.7	0.9	6.3	0.7	5.4
6			1.7	9.7	1.2	6.9	0.9	5.6
7			2.2	10.7	1.5	7.5	1.2	6.0
8			2.7	11.7	1.9	8.1	1.4	6.4
9			3.3	12.7	2.2	8.8	1.7	6.8
10			3.8	13.6	2.6	9.4	2.0	7.3
11			4.3	14.6	3.0	10.1	2.3	7.7
12			4.9	15.5	3.4	10.7	2.6	8.2
13			5.5	16.4	3.7	11.3	2.8	8.7
14			6.1	17.3	4.1	12.0	3.1	9.2
15			6.7	18.2	4.5	12.6	3.5	9.6
16			7.3	19.1	4.9	13.2	3.8	10.1
17			7.9	20.0	5.4	13.8	4.1	10.6
18			8.5	20.9	5.8	14.5	4.4	11.0
19			9.1	21.8	6.2	15.1	4.7	11.5
20			9.7	22.7	6.6	15.7	5.0	12.0
21			10.4	23.5	7.0	16.3	5.3	12.5
22			11.0	24.4	7.5	16.9	5.7	12.9
23			11.6	25.3	7.9	17.5	6.0	13.4
24			12.3	26.1	8.3	18.1	6.3	13.8
25			12.9	27.0	8.8	18.7	6.7	14.3

*The two-step (double) sample confidence limits in this table are based on an initial sample of size n_1 with an additional sample of size n_2 *only* if occurrences are found in the initial sample.

Table A-2

Two-Step Sampling Confidence Limits*

Initial Sample (n_1) = 70

Confidence Level = 95% (Two-Sided), 97.5% (One-Sided)

N = *10,000*

Occurrences in Sample	Single Sample n_1		Two-Step (Double) Sample ($n_1 + n_2$)					
			$n_2 = 60$		$n_2 = 120$		$n_2 = 180$	
	LL	UL	LL	UL	LL	UL	LL	UL
0	0.0%	4.2%	—	—	—	—	—	—
1			0.1%	5.3%	0.1%	5.1%	0.1%	5.1%
2			0.2	5.9	0.2	5.2	0.1	5.1
3			0.5	6.8	0.4	5.4	0.3	5.1
4			0.9	7.7	0.6	5.8	0.5	5.2
5			1.3	8.7	0.9	6.3	0.7	5.4
6			1.7	9.8	1.2	6.9	0.9	5.7
7			2.2	10.8	1.5	7.5	1.2	6.0
8			2.7	11.7	1.9	8.2	1.4	6.4
9			3.2	12.7	2.2	8.8	1.7	6.9
10			3.8	13.7	2.6	9.4	2.0	7.3
11			4.3	14.6	2.9	10.1	2.2	7.8
12			4.9	15.5	3.3	10.7	2.5	8.2
13			5.5	16.4	3.7	11.4	2.8	8.7
14			6.0	17.3	4.1	12.0	3.1	9.2
15			6.6	18.3	4.5	12.6	3.4	9.7
16			7.2	19.2	4.9	13.3	3.7	10.2
17			7.8	20.1	5.3	13.9	4.0	10.6
18			8.4	21.0	5.7	14.5	4.4	11.1
19			9.1	21.8	6.2	15.1	4.7	11.6
20			9.7	22.7	6.6	15.7	5.0	12.0
21			10.3	23.6	7.0	16.3	5.3	12.5
22			11.0	24.4	7.4	17.0	5.6	13.0
23			11.6	25.3	7.9	17.6	6.0	13.4
24			12.2	26.2	8.3	18.2	6.3	13.9
25			12.9	27.0	8.8	18.8	6.6	14.4

*The two-step (double) sample confidence limits in this table are based on an initial sample of size n_1 with an additional sample of size n_2 *only* if occurrences are found in the initial sample.

Table A-2

Two-Step Sampling Confidence Limits*

Initial Sample (n_1) = ***80***

Confidence Level = 95% (Two-Sided), 97.5% (One-Sided)

N = ***1000***

Occurrences in Sample	Single Sample n_1		Two-Step (Double) Sample ($n_1 + n_2$)					
			$n_2 = 60$		$n_2 = 120$		$n_2 = 180$	
	LL	UL	LL	UL	LL	UL	LL	UL
0	0.0%	3.5%	—	—	—	—	—	—
1			0.1%	4.5%	0.1%	4.3%	0.1%	4.3%
2			0.2	5.1	0.2	4.4	0.2	4.3
3			0.5	6.0	0.4	4.6	0.3	4.3
4			0.9	6.9	0.7	5.1	0.5	4.4
5			1.3	7.8	1.0	5.6	0.8	4.6
6			1.7	8.8	1.3	6.2	1.0	4.9
7			2.2	9.7	1.6	6.8	1.3	5.3
8			2.7	10.6	1.9	7.4	1.5	5.7
9			3.2	11.5	2.3	8.0	1.8	6.2
10			3.7	12.3	2.6	8.6	2.1	6.6
11			4.2	13.2	3.0	9.2	2.4	7.0
12			4.8	14.1	3.4	9.8	2.7	7.5
13			5.3	14.9	3.8	10.4	3.0	8.0
14			5.9	15.8	4.2	11.0	3.3	8.4
15			6.4	16.6	4.6	11.6	3.6	8.9
16			7.0	17.4	5.0	12.2	3.9	9.3
17			7.5	18.3	5.3	12.8	4.2	9.8
18			8.1	19.1	5.8	13.4	4.5	10.2
19			8.7	19.9	6.2	13.9	4.8	10.6
20			9.3	20.7	6.6	14.5	5.1	11.1
21			9.9	21.5	7.0	15.1	5.4	11.5
22			10.5	22.3	7.4	15.6	5.8	12.0
23			11.1	23.1	7.8	16.2	6.1	12.4
24			11.7	23.9	8.2	16.8	6.4	12.9
25			12.3	24.7	8.7	17.3	6.7	13.3

*The two-step (double) sample confidence limits in this table are based on an initial sample of size n_1 with an additional sample of size n_2 *only* if occurrences are found in the initial sample.

Table A-2
Two-Step Sampling Confidence Limits*
Initial Sample (n_1) = 80
Confidence Level = 95% (Two-Sided), 97.5% (One-Sided)
N = 5000

Occurrences in Sample	Single Sample n_1		Two-Step (Double) Sample ($n_1 + n_2$)					
			n_2 = 60		n_2 = 120		n_2 = 180	
	LL	UL	LL	UL	LL	UL	LL	UL
0	0.0%	3.6%	—	—	—	—	—	—
1			0.1%	4.7%	0.1%	4.5%	0.1%	4.5%
2			0.2	5.3	0.2	4.6	0.1	4.5
3			0.5	6.2	0.3	4.9	0.3	4.5
4			0.8	7.1	0.6	5.3	0.5	4.7
5			1.2	8.1	0.9	5.9	0.7	4.9
6			1.6	9.0	1.1	6.5	0.9	5.2
7			2.1	10.0	1.5	7.1	1.1	5.6
8			2.5	10.9	1.8	7.7	1.4	6.0
9			3.0	11.8	2.1	8.3	1.6	6.5
10			3.5	12.7	2.5	8.9	1.9	6.9
11			4.0	13.5	2.8	9.6	2.2	7.4
12			4.6	14.4	3.2	10.2	2.5	7.9
13			5.1	15.2	3.6	10.8	2.7	8.3
14			5.6	16.1	3.9	11.4	3.0	8.8
15			6.2	17.0	4.3	12.0	3.3	9.3
16			6.7	17.8	4.7	12.6	3.6	9.7
17			7.3	18.6	5.1	13.2	3.9	10.2
18			7.9	19.5	5.5	13.8	4.2	10.6
19			8.4	20.3	5.9	14.3	4.5	11.1
20			9.0	21.1	6.3	14.9	4.8	11.5
21			9.6	21.9	6.7	15.5	5.1	12.0
22			10.2	22.7	7.1	16.1	5.4	12.4
23			10.8	23.5	7.5	16.7	5.8	12.9
24			11.4	24.3	7.9	17.2	6.1	13.3
25			12.0	25.1	8.3	17.8	6.4	13.8

*The two-step (double) sample confidence limits in this table are based on an initial sample of size n_1 with an additional sample of size n_2 *only* if occurrences are found in the initial sample.

Table A-2
*Two-Step Sampling Confidence Limits**
Initial Sample (n_1) = *80*
Confidence Level = 95% (Two-Sided), 97.5% (One-Sided)
N = *10,000*

Occurrences in Sample	Single Sample n_1		Two-Step (Double) Sample ($n_1 + n_2$)					
			n_2 = 60		n_2 = 120		n_2 = 180	
	LL	UL	LL	UL	LL	UL	LL	UL
0	0.0%	3.7%	—	—	—	—	—	—
1			0.1%	4.7%	0.1%	4.5%	0.1%	4.5%
2			0.2	5.4	0.1	4.6	0.1	4.5
3			0.5	6.2	0.3	4.9	0.3	4.6
4			0.8	7.2	0.6	5.3	0.5	4.7
5			1.2	8.1	0.8	5.9	0.7	4.9
6			1.6	9.1	1.1	6.5	0.9	5.3
7			2.0	10.0	1.4	7.1	1.1	5.7
8			2.5	10.9	1.8	7.7	1.4	6.1
9			3.0	11.8	2.1	8.4	1.6	6.5
10			3.5	12.7	2.4	9.0	1.9	7.0
11			4.0	13.6	2.8	9.6	2.2	7.4
12			4.5	14.5	3.2	10.2	2.4	7.9
13			5.1	15.3	3.5	10.8	2.7	8.4
14			5.6	16.1	3.9	11.4	3.0	8.8
15			6.1	17.0	4.3	12.0	3.3	9.3
16			6.7	17.9	4.7	12.6	3.6	9.7
17			7.3	18.7	5.1	13.2	3.9	10.2
18			7.8	19.5	5.5	13.8	4.2	10.6
19			8.4	20.3	5.9	14.3	4.5	11.1
20			9.0	21.2	6.3	14.9	4.8	11.5
21			9.6	22.0	6.7	15.5	5.1	12.0
22			10.2	22.8	7.1	16.1	5.4	12.4
23			10.7	23.6	7.5	16.7	5.8	12.9
24			11.3	24.4	7.9	17.2	6.1	13.3
25			11.9	25.2	8.3	17.8	6.4	13.8

*The two-step (double) sample confidence limits in this table are based on an initial sample of size n_1 with an additional sample of size n_2 *only* if occurrences are found in the initial sample.

Table A-2

Two-Step Sampling Confidence Limits*

Initial Sample* (n_1) *= 90

Confidence Level = 95% (Two-Sided), 97.5% (One-Sided)

N = *1000*

Occurrences in Sample	Single Sample n_1		Two-Step (Double) Sample ($n_1 + n_2$)					
			$n_2 = 60$		$n_2 = 120$		$n_2 = 180$	
	LL	UL	LL	UL	LL	UL	LL	UL
0	0.0%	3.1%	—	—	—	—	—	—
1			0.1%	4.0%	0.1%	3.8%	0.1%	3.8%
2			0.2	4.7	0.2	3.9	0.2	3.8
3			0.5	5.5	0.3	4.2	0.3	3.9
4			0.8	6.4	0.6	4.7	0.5	4.0
5			1.2	7.3	0.9	5.2	0.7	4.3
6			1.6	8.2	1.2	5.8	1.0	4.6
7			2.1	9.0	1.5	6.4	1.2	5.0
8			2.5	9.9	1.9	7.0	1.5	5.4
9			3.0	10.7	2.2	7.6	1.8	5.9
10			3.5	11.5	2.5	8.2	2.0	6.3
11			4.0	12.3	2.9	8.8	2.3	6.8
12			4.5	13.1	3.2	9.3	2.6	7.2
13			5.0	13.7	3.6	9.9	2.9	7.6
14			5.5	14.7	4.0	10.5	3.2	8.1
15			6.0	15.5	4.3	11.0	3.4	8.5
16			6.5	16.3	4.7	11.6	3.7	8.9
17			7.1	17.1	5.1	12.2	4.0	9.4
18			7.6	17.8	5.5	12.7	4.3	9.8
19			8.1	18.6	5.9	13.3	4.6	10.2
20			8.7	19.3	6.3	13.6	4.9	10.7
21			9.2	20.1	6.7	14.3	5.3	11.1
22			9.8	20.9	7.1	14.9	5.6	11.5
23			10.3	21.6	7.5	15.4	5.9	11.9
24			10.9	22.3	7.9	16.0	6.2	12.4
25			11.5	23.1	8.3	16.5	6.5	12.8

*The two-step (double) sample confidence limits in this table are based on an initial sample of size n_1 with an additional sample of size n_2 *only* if occurrences are found in the initial sample.

Table A-2
*Two-Step Sampling Confidence Limits**
Initial Sample (n_1) = *90*
Confidence Level = 95% (Two-Sided), 97.5% (One-Sided)
N = *5000*

Occurrences in Sample	Single Sample n_1		Two-Step (Double) Sample ($n_1 + n_2$)					
			$n_2 = 60$		$n_2 = 120$		$n_2 = 180$	
	LL	UL	LL	UL	LL	UL	LL	UL
0	0.0%	3.2%	—	—	—	—	—	—
1			0.1%	4.2%	0.1%	4.0%	0.1%	4.0%
2			0.2	4.9	0.1	4.2	0.1	4.0
3			0.4	5.8	0.3	4.5	0.3	4.1
4			0.8	6.7	0.6	5.0	0.4	4.3
5			1.1	7.6	0.8	5.5	0.6	4.6
6			1.5	8.4	1.1	6.1	0.9	4.9
7			1.9	9.3	1.4	6.7	1.1	5.3
8			2.4	10.2	1.7	7.3	1.3	5.8
9			2.8	11.0	2.0	7.9	1.6	6.2
10			3.3	11.8	2.3	8.5	1.8	6.7
11			3.8	12.7	2.7	9.1	2.1	7.1
12			4.2	13.5	3.0	9.7	2.4	7.6
13			4.7	14.3	3.4	10.3	2.6	8.0
14			5.2	15.1	3.7	10.8	2.9	8.5
15			5.8	15.9	4.1	11.4	3.2	8.9
16			6.3	16.7	4.5	12.0	3.5	9.4
17			6.8	17.4	4.8	12.6	3.8	9.8
18			7.3	18.2	5.2	13.1	4.1	10.2
19			7.9	19.0	5.6	13.7	4.4	10.7
20			8.4	19.7	6.0	14.2	4.6	11.1
21			8.9	20.5	6.4	14.8	4.9	11.5
22			9.5	21.3	6.8	15.3	5.2	12.0
23			10.0	22.0	7.1	15.9	5.5	12.4
24			10.6	22.8	7.5	16.0	5.9	12.8
25			11.2	23.5	7.9	17.0	6.2	13.3

*The two-step (double) sample confidence limits in this table are based on an initial sample of size n_1 with an additional sample of size n_2 *only* if occurrences are found in the initial sample.

Table A-2
Two-Step Sampling Confidence Limits*
Initial Sample (n_1) = _90_
Confidence Level = 95% (Two-Sided), 97.5% (One-Sided)
N = **_10,000_**

Occurrences in Sample	Single Sample n_1		Two-Step (Double) Sample ($n_1 + n_2$)					
			$n_2 = 60$		$n_2 = 120$		$n_2 = 180$	
	LL	UL	LL	UL	LL	UL	LL	UL
0	0.0%	3.3%	—	—	—	—	—	—
1			0.1%	4.3%	0.1%	4.0%	0.1%	4.0%
2			0.2	4.9	0.1	4.2	0.1	4.0
3			0.4	5.8	0.3	4.5	0.3	4.1
4			0.7	6.7	0.5	5.0	0.4	4.3
5			1.1	7.6	0.8	5.6	0.6	4.6
6			1.5	8.5	1.1	6.1	0.8	5.0
7			1.9	9.3	1.4	6.7	1.1	5.4
8			2.3	10.2	1.7	7.3	1.3	5.8
9			2.8	11.0	2.0	8.0	1.6	6.3
10			3.3	11.9	2.3	8.5	1.8	6.7
11			3.7	12.7	2.7	9.1	2.1	7.2
12			4.2	13.5	3.0	9.7	2.3	7.6
13			4.7	14.3	3.4	10.3	2.6	8.1
14			5.2	15.1	3.7	10.9	2.9	8.5
15			5.7	15.9	4.1	11.5	3.2	9.0
16			6.2	16.7	4.4	12.0	3.5	9.4
17			6.8	17.5	4.8	12.6	3.7	9.8
18			7.3	18.3	5.2	13.2	4.0	10.3
19			7.8	19.0	5.6	13.7	4.3	10.7
20			8.4	19.8	5.9	14.3	4.6	11.2
21			8.9	20.6	6.3	14.8	4.9	11.6
22			9.5	21.3	6.7	15.4	5.2	12.0
23			10.0	22.1	7.1	15.9	5.5	12.5
24			10.6	22.8	7.5	16.5	5.8	12.9
25			11.1	23.6	7.9	17.0	6.1	13.3

*The two-step (double) sample confidence limits in this table are based on an initial sample of size n_1 with an additional sample of size n_2 *only* if occurrences are found in the initial sample.

Table A-2
Two-Step Sampling Confidence Limits*
Initial Sample (n_1) = 100
Confidence Level = 95% (Two-Sided), 97.5% (One-Sided)
N = 1000

Occurrences in Sample	Single Sample n_1		Two-Step (Double) Sample ($n_1 + n_2$)					
			$n_2 = 60$		$n_2 = 120$		$n_2 = 180$	
	LL	UL	LL	UL	LL	UL	LL	UL
0	0.0%	2.8%	—	—	—	—	—	—
1			0.1%	3.7%	0.1%	3.4%	0.1%	3.4%
2			0.2	4.3	0.2	3.6	0.2	3.4
3			0.5	5.1	0.3	3.9	0.3	3.5
4			0.8	6.0	0.6	4.4	0.5	3.7
5			1.1	6.8	0.9	5.0	0.7	4.0
6			1.5	7.6	1.2	5.5	1.0	4.4
7			1.9	8.4	1.5	6.1	1.2	4.8
8			2.4	9.2	1.8	6.7	1.4	5.2
9			2.8	10.0	2.1	7.2	1.7	5.6
10			3.3	10.8	2.4	7.8	2.0	6.1
11			3.7	11.6	2.8	8.3	2.2	6.5
12			4.2	12.3	3.1	8.9	2.5	6.9
13			4.7	13.1	3.5	9.4	2.8	7.3
14			5.1	13.8	3.8	10.0	3.1	7.8
15			5.6	14.5	4.2	10.5	3.3	8.2
16			6.1	15.3	4.5	11.1	3.6	8.6
17			6.6	16.0	4.9	11.6	3.9	9.0
18			7.1	16.7	5.3	12.1	4.2	9.4
19			7.6	17.4	5.6	12.6	4.5	9.9
20			8.2	18.1	6.0	13.2	4.8	10.3
21			8.7	18.9	6.4	13.7	5.1	10.7
22			9.2	19.6	6.8	14.2	5.4	11.1
23			9.7	20.3	7.1	14.7	5.7	11.5
24			10.2	21.0	7.5	15.2	6.0	11.9
25			10.8	21.7	7.9	15.8	6.3	12.3

*The two-step (double) sample confidence limits in this table are based on an initial sample of size n_1 with an additional sample of size n_2 *only* if occurrences are found in the initial sample.

Table A-2
Two-Step Sampling Confidence Limits*
Initial Sample (n_1) ***= 100***
Confidence Level = 95% (Two-Sided), 97.5% (One-Sided)
N = ***5000***

Occurrences in Sample	Single Sample n_1		Two-Step (Double) Sample $(n_1 + n_2)$					
			$n_2 = 60$		$n_2 = 120$		$n_2 = 180$	
	LL	UL	LL	UL	LL	UL	LL	UL
0	0.0%	2.9%	—	—	—	—	—	—
1			0.1%	3.9%	0.1%	3.6%	0.1%	3.6%
2			0.2	4.6	0.1	3.8	0.1	3.6
3			0.4	5.4	0.3	4.2	0.3	3.7
4			0.7	6.2	0.5	4.7	0.4	4.0
5			1.0	7.1	0.8	5.2	0.6	4.3
6			1.4	7.9	1.0	5.8	0.8	4.7
7			1.8	8.7	1.3	6.4	1.0	5.1
8			2.2	9.5	1.6	7.0	1.3	5.5
9			2.6	10.3	1.9	7.6	1.5	6.0
10			3.1	11.1	2.2	8.1	1.8	6.4
11			3.5	11.9	2.6	8.7	2.0	6.9
12			4.0	12.7	2.9	9.3	2.3	7.3
13			4.4	13.4	3.2	9.8	2.5	7.7
14			4.9	14.2	3.6	10.4	2.8	8.2
15			5.4	14.9	3.9	10.9	3.1	8.6
16			5.9	15.6	4.3	11.4	3.4	9.0
17			6.4	16.4	4.6	12.0	3.6	9.4
18			6.9	17.1	5.0	12.5	3.9	9.9
19			7.4	17.8	5.3	13.1	4.2	10.3
20			7.9	18.5	5.7	13.6	4.5	10.7
21			8.4	19.3	6.1	14.1	4.8	11.1
22			8.9	20.0	6.4	14.6	5.1	11.5
23			9.4	20.7	6.8	15.2	5.4	12.0
24			9.9	21.4	7.2	15.7	5.6	12.4
25			10.4	22.1	7.6	16.2	5.9	12.8

*The two-step (double) sample confidence limits in this table are based on an initial sample of size n_1 with an additional sample of size n_2 *only* if occurrences are found in the initial sample.

Table A-2
*Two-Step Sampling Confidence Limits**
Initial Sample (n_1) = *100*
Confidence Level = 95% (Two-Sided), 97.5% (One-Sided)
N = *10,000*

Occurrences in Sample	Single Sample n_1		Two-Step (Double) Sample ($n_1 + n_2$)					
			n_2 = 60		n_2 = 120		n_2 = 180	
	LL	UL	LL	UL	LL	UL	LL	UL
0	0.0%	2.9%	—	—	—	—	—	—
1			0.1%	3.9%	0.1%	3.6%	0.1%	3.6%
2			0.2	4.6	0.1	3.8	0.1	3.6
3			0.4	5.4	0.3	4.2	0.2	3.8
4			0.7	6.3	0.5	4.7	0.4	4.0
5			1.0	7.1	0.8	5.3	0.6	4.3
6			1.4	8.0	1.0	5.8	0.8	4.7
7			1.8	8.8	1.3	6.4	1.0	5.1
8			2.2	9.6	1.6	7.0	1.3	5.6
9			2.6	10.4	1.9	7.6	1.5	6.0
10			3.1	11.2	2.2	8.2	1.7	6.5
11			3.5	11.9	2.5	8.7	2.0	6.9
12			4.0	12.7	2.9	9.3	2.3	7.3
13			4.4	13.5	3.2	9.9	2.5	7.8
14			4.9	14.2	3.5	10.4	2.8	8.2
15			5.4	14.9	3.9	11.0	3.1	8.6
16			5.9	15.7	4.2	11.5	3.3	9.1
17			6.3	16.4	4.6	12.0	3.6	9.5
18			6.8	17.2	5.0	12.6	3.9	9.9
19			7.3	17.9	5.3	13.1	4.2	10.3
20			7.8	18.6	5.7	13.6	4.4	10.8
21			8.3	19.3	6.0	14.2	4.7	11.2
22			8.9	20.0	6.4	14.7	5.0	11.6
23			9.4	20.7	6.8	15.2	5.3	12.0
24			9.9	21.4	7.2	15.7	5.6	12.4
25			10.4	22.1	7.5	16.3	5.9	12.8

*The two-step (double) sample confidence limits in this table are based on an initial sample of size n_1 with an additional sample of size n_2 *only* if occurrences are found in the initial sample.

Table A-2
Two-Step Sampling Confidence Limits*
Initial Sample (n_1) = 50
Confidence Level = 99% (Two-Sided), 99.5% (One-Sided)
N = **1000**

Occurrences in Sample	Single Sample n_1		Two-Step (Double) Sample ($n_1 + n_2$)					
			n_2 = 60		n_2 = 120		n_2 = 180	
	LL	UL	LL	UL	LL	UL	LL	UL
0	0.0%	8.6%	—	—	—	—	—	—
1			0.1%	9.8%	0.1%	9.8%	0.1%	9.8%
2			0.2	9.9	0.2	9.8	0.2	9.8
3			0.4	10.3	0.3	9.8	0.3	9.8
4			0.8	11.1	0.5	9.8	0.4	9.8
5			1.2	12.2	0.8	9.9	0.7	9.8
6			1.6	13.3	1.1	10.0	0.9	9.8
7			2.1	14.5	1.4	10.3	1.1	9.8
8			2.6	15.6	1.8	10.7	1.4	9.8
9			3.2	16.7	2.1	11.1	1.7	9.9
10			3.7	17.8	2.5	11.9	1.9	9.9
11			4.3	19.0	2.9	12.5	2.2	10.1
12			4.9	20.0	3.3	13.2	2.5	10.4
13			5.6	21.2	3.7	13.9	2.8	10.7
14			6.2	22.3	4.1	14.6	3.1	11.1
15			6.8	23.5	4.5	15.3	3.4	11.5
16			7.5	24.6	4.9	16.0	3.7	12.0
17			8.2	25.6	5.4	16.7	4.1	12.5
18			8.9	26.7	5.8	17.4	4.4	13.0
19			9.5	27.8	6.3	18.1	4.7	13.4
20			10.2	28.8	6.7	18.8	5.1	13.9
21			10.9	29.8	7.2	19.5	5.4	14.4
22			11.7	30.8	7.6	20.2	5.7	14.9
23			12.4	31.8	8.1	20.9	6.1	15.5
24			13.1	32.8	8.5	21.6	6.4	16.0
25			13.8	33.8	9.0	22.3	6.8	16.5

*The two-step (double) sample confidence limits in this table are based on an initial sample of size n_1 with an additional sample of size n_2 *only* if occurrences are found in the initial sample.

Table A-2

Two-Step Sampling Confidence Limits*

Initial Sample **(n_1) = 50**

Confidence Level = 99% (Two-Sided), 99.5% (One-Sided)

N = 5000

Occurrences in Sample	Single Sample n_1		Two-Step (Double) Sample ($n_1 + n_2$)					
			n_2 = 60		n_2 = 120		n_2 = 180	
	LL	UL	LL	UL	LL	UL	LL	UL
0	0.0%	8.7%	—	—	—	—	—	—
1			0.1%	10.0%	0.1%	10.0%	0.1%	10.0%
2			0.1	10.2	0.1	10.0	0.1	10.0
3			0.3	10.6	0.2	10.0	0.2	10.0
4			0.7	11.5	0.4	10.0	0.3	10.0
5			1.0	12.5	0.7	10.0	0.5	10.0
6			1.5	13.7	1.0	10.3	0.7	10.0
7			1.9	14.8	1.3	10.6	1.0	10.0
8			2.4	16.0	1.6	11.1	1.2	10.0
9			3.0	17.1	1.9	11.7	1.4	10.1
10			3.5	18.3	2.3	12.3	1.7	10.3
11			4.1	19.4	2.6	13.0	2.0	10.5
12			4.7	20.6	3.0	13.7	2.2	10.8
13			5.3	21.7	3.4	14.4	2.5	11.2
14			5.9	22.9	3.8	15.1	2.8	11.6
15			6.5	24.0	4.2	15.8	3.1	12.0
16			7.2	25.1	4.6	16.6	3.4	12.5
17			7.9	26.1	5.0	17.3	3.7	13.0
18			8.5	27.2	5.5	18.0	4.0	13.5
19			9.2	28.2	5.9	18.7	4.4	14.0
20			9.9	29.3	6.3	19.4	4.7	14.5
21			10.6	30.4	6.8	20.1	5.0	15.6
22			11.3	31.5	7.2	20.8	5.4	16.2
23			12.0	32.6	7.7	21.5	6.0	16.7
24			12.7	33.7	8.1	22.2	6.3	17.2
25			13.5	34.8	8.6	22.8	6.7	17.7

*The two-step (double) sample confidence limits in this table are based on an initial sample of size n_1 with an additional sample of size n_2 *only* if occurrences are found in the initial sample.

Table A-2

Two-Step Sampling Confidence Limits*

***Initial Sample* (n_1) = 50**

Confidence Level = 99% (Two-Sided), 99.5% (One-Sided)

N = ***10,000***

Occurrences in Sample	Single Sample n_1		Two-Step (Double) Sample (n_1 + n_2)					
			n_2 = 60		n_2 = 120		n_2 = 180	
	LL	UL	LL	UL	LL	UL	LL	UL
0	0.0%	8.8%	—	—	—	—	—	—
1			0.1%	10.0%	0.1%	10.0%	0.1%	10.0%
2			0.1	10.2	0.1	10.0	0.1	10.0
3			0.3	10.7	0.2	10.0	0.2	10.0
4			0.6	11.5	0.4	10.1	0.3	10.0
5			1.0	12.5	0.7	10.1	0.5	10.0
6			1.4	13.7	0.9	10.3	0.7	10.0
7			1.9	14.9	1.2	10.7	0.9	10.0
8			2.4	16.0	1.6	11.1	1.2	10.1
9			2.9	17.1	1.9	11.7	1.4	10.1
10			3.5	18.3	2.3	12.4	1.7	10.3
11			4.1	19.4	2.6	13.0	1.9	10.5
12			4.7	20.5	3.0	13.8	2.2	10.8
13			5.3	21.7	3.4	14.5	2.5	11.2
14			5.9	22.8	3.8	15.2	2.8	11.6
15			6.5	24.0	4.2	15.9	3.1	12.1
16			7.2	25.1	4.6	16.6	3.4	12.6
17			7.8	26.2	5.0	17.3	3.7	13.1
18			8.5	27.3	5.4	18.0	4.0	13.6
19			9.2	28.3	5.9	18.7	4.3	14.1
20			9.9	29.4	6.3	19.5	4.6	14.6
21			10.5	30.5	6.7	20.2	5.0	15.1
22			11.3	31.6	7.2	20.8	5.3	15.6
23			12.0	32.7	7.6	21.5	5.6	16.1
24			12.7	33.8	8.1	22.2	5.9	16.6
25			13.4	34.9	8.5	22.9	6.3	17.1

*The two-step (double) sample confidence limits in this table are based on an initial sample of size n_1 with an additional sample of size n_2 *only* if occurrences are found in the initial sample.

Table A-2
Two-Step Sampling Confidence Limits*
Initia; Sample (n_1) = 60
Confidence Level = 99% (Two-Sided), 99.5% (One-Sided)
N = 1000

Occurrences in Sample	Single Sample n_1		Two-Step (Double) Sample ($n_1 + n_2$)					
			$n_2 = 60$		$n_2 = 120$		$n_2 = 180$	
	LL	UL	LL	UL	LL	UL	LL	UL
0	0.0%	7.2%	—	—	—	—	—	—
1			0.1%	8.2%	0.1%	8.2%	0.1%	8.2%
2			0.2	8.4	0.2	8.2	0.2	8.2
3			0.3	9.0	0.3	8.2	0.3	8.2
4			0.7	10.0	0.5	8.3	0.4	8.2
5			1.1	11.0	0.8	8.4	0.6	8.2
6			1.5	12.1	1.1	8.7	0.9	8.2
7			1.9	13.2	1.4	9.2	1.1	8.2
8			2.4	14.3	1.7	9.8	1.3	8.3
9			2.9	15.3	2.0	10.4	1.6	8.5
10			3.4	16.4	2.4	11.1	1.9	8.7
11			4.0	17.4	2.7	11.7	2.1	9.1
12			4.5	18.5	3.1	12.4	2.4	9.5
13			5.1	19.5	3.5	13.1	2.7	9.9
14			5.7	20.6	3.9	13.8	3.0	10.4
15			6.3	21.6	4.3	14.5	3.3	10.9
16			6.9	22.6	4.7	15.1	3.6	11.3
17			7.5	23.6	5.1	15.8	3.9	11.8
18			8.1	24.6	5.5	16.5	4.2	12.3
19			8.8	25.5	5.9	17.1	4.5	12.8
20			9.4	26.5	6.3	17.8	4.9	13.3
21			10.0	27.4	6.8	18.4	5.2	13.8
22			10.7	28.4	7.2	19.1	5.5	14.3
23			11.4	29.3	7.6	19.7	5.8	14.8
24			12.0	30.2	8.1	20.4	6.2	15.3
25			12.7	31.1	8.5	21.0	6.5	15.7

*The two-step (double) sample confidence limits in this table are based on an initial sample of size n_1 with an additional sample of size n_2 *only* if occurrences are found in the initial sample.

Table A-2
*Two-Step Sampling Confidence Limits**
Initial Sample (n_1) *= 60*
Confidence Level = 99% (Two-Sided), 99.5% (One-Sided)
N = *5000*

Occurrences in Sample	Single Sample n_1		Two-Step (Double) Sample ($n_1 + n_2$)					
			n_2 = 60		n_2 = 120		n_2 = 180	
	LL	UL	LL	UL	LL	UL	LL	UL
0	0.0%	7.3%	—	—	—	—	—	—
1			0.1%	8.4%	0.1%	8.4%	0.1%	8.4%
2			0.1	8.7	0.1	8.4	0.1	8.4
3			0.3	9.4	0.2	8.4	0.2	8.4
4			0.6	10.3	0.4	8.5	0.3	8.4
5			0.9	11.4	0.6	8.7	0.5	8.4
6			1.3	12.5	0.9	9.1	0.7	8.4
7			1.8	13.6	1.2	9.6	0.9	8.5
8			2.2	14.7	1.5	10.2	1.1	8.6
9			2.7	15.8	1.8	10.8	1.4	8.8
10			3.2	16.8	2.1	11.5	1.6	9.2
11			3.7	17.9	2.5	12.2	1.9	9.9
12			4.3	18.9	2.9	12.9	2.1	10.0
13			4.8	20.0	3.2	13.6	2.4	10.4
14			5.4	21.0	3.6	14.3	2.7	10.9
15			6.0	22.1	4.0	15.0	3.0	11.4
16			6.6	23.1	4.4	15.7	3.3	11.9
17			7.2	24.1	4.8	16.3	3.6	12.4
18			7.8	25.1	5.2	17.0	3.9	12.9
19			8.4	26.0	5.6	17.7	4.2	13.4
20			9.0	27.0	6.0	18.3	4.5	13.9
21			9.7	28.0	6.4	19.0	4.8	14.4
22			10.3	28.9	6.8	19.7	5.1	14.9
23			11.0	29.9	7.2	20.3	5.4	15.4
24			11.6	30.9	7.7	21.0	5.7	15.9
25			12.3	31.9	8.1	22.6	6.1	16.4

*The two-step (double) sample confidence limits in this table are based on an initial sample of size n_1 with an additional sample of size n_2 *only* if occurrences are found in the initial sample.

Table A-2
Two-Step Sampling Confidence Limits*
Initial Sample (n_1) = 60
Confidence Level = 99% (Two-Sided), 99.5% (One-Sided)
N = 10,000

Occurrences in Sample	Single Sample n_1		Two-Step (Double) Sample ($n_1 + n_2$)					
			$n_2 = 60$		$n_2 = 120$		$n_2 = 180$	
	LL	UL	LL	UL	LL	UL	LL	UL
0	0.0%	7.4%	—	—	—	—	—	—
1			0.1%	8.5%	0.1%	8.4%	0.1%	8.4%
2			0.1	8.7	0.1	8.4	0.1	8.4
3			0.3	9.4	0.2	8.4	0.2	8.4
4			0.6	10.3	0.4	8.5	0.3	8.4
5			0.9	11.4	0.6	8.7	0.5	8.4
6			1.3	12.5	0.9	9.1	0.7	8.5
7			1.7	13.7	1.2	9.6	0.9	8.5
8			2.2	14.8	1.5	10.2	1.1	8.7
9			2.7	15.8	1.8	10.9	1.4	8.9
10			3.2	16.9	2.1	11.6	1.6	9.2
11			3.7	17.9	2.5	12.2	1.9	9.6
12			4.3	19.0	2.8	12.9	2.1	10.0
13			4.8	20.0	3.2	13.6	2.4	10.5
14			5.4	21.1	3.6	14.3	2.7	11.0
15			6.0	22.1	3.9	15.0	3.0	11.4
16			6.5	23.1	4.3	15.7	3.2	11.9
17			7.1	24.1	4.7	16.4	3.5	12.4
18			7.8	25.1	5.1	17.1	3.8	12.9
19			8.4	26.1	5.5	17.7	4.1	13.5
20			9.0	27.1	5.9	18.4	4.4	14.0
21			9.6	28.0	6.4	19.1	4.7	14.5
22			10.3	29.0	6.8	19.7	5.1	15.0
23			10.9	29.9	7.2	20.4	5.4	15.5
24			11.6	30.9	7.6	21.0	5.7	15.9
25			12.2	31.9	8.1	22.8	6.0	16.4

*The two-step (double) sample confidence limits in this table are based on an initial sample of size n_1 with an additional sample of size n_2 *only* if occurrences are found in the initial sample.

Table A-2
Two-Step Sampling Confidence Limits*
Initial Sample (n_1) = 70
Confidence Level = 99% (Two-Sided), 99.5% (One-Sided)
N = **1000**

Occurrences in Sample	Single Sample n_1		Two-Step (Double) Sample ($n_1 + n_2$)					
			$n_2 = 60$		$n_2 = 120$		$n_2 = 180$	
	LL	UL	LL	UL	LL	UL	LL	UL
0	0.0%	6.1%	—	—	—	—	—	—
1			0.1%	7.1%	0.1%	7.0%	0.1%	7.0%
2			0.2	7.4	0.2	7.0	0.2	7.0
3			0.3	8.1	0.3	7.1	0.3	7.0
4			0.7	9.1	0.5	7.2	0.4	7.0
5			1.0	10.1	0.7	7.5	0.6	7.0
6			1.4	11.2	1.0	7.9	0.8	7.1
7			1.8	12.2	1.3	8.5	1.1	7.2
8			2.2	13.2	1.6	9.1	1.3	7.4
9			2.7	14.2	1.9	9.8	1.5	7.7
10			3.2	15.2	2.3	10.4	1.8	8.1
11			3.7	16.1	2.6	11.1	2.1	8.5
12			4.2	17.1	3.0	11.7	2.3	8.9
13			4.7	18.0	3.3	12.4	2.6	9.4
14			5.3	19.0	3.7	13.0	2.9	9.9
15			5.8	20.0	4.1	13.7	3.2	10.4
16			6.4	20.9	4.5	14.3	3.5	10.8
17			6.9	21.8	4.8	15.0	3.8	11.3
18			7.5	22.7	5.2	15.6	4.1	11.8
19			8.1	23.6	5.6	16.2	4.4	12.3
20			8.7	24.5	6.0	16.9	4.7	12.7
21			9.3	25.4	6.4	17.5	5.0	13.2
22			9.9	26.3	6.8	18.1	5.3	13.7
23			10.5	27.2	7.3	18.7	5.6	14.2
24			11.1	28.0	7.7	19.3	5.9	14.6
25			11.7	28.9	8.1	19.9	6.2	15.1

*The two-step (double) sample confidence limits in this table are based on an initial sample of size n_1 with an additional sample of size n_2 *only* if occurrences are found in the initial sample.

Table A-2
Two-Step Sampling Confidence Limits*
Initial Sample (n_1) ***= 70***
Confidence Level = 99% (Two-Sided), 99.5% (One-Sided)
N = ***5000***

Occurrences in Sample	Single Sample n_1		Two-Step (Double) Sample $(n_1 + n_2)$					
			$n_2 = 60$		$n_2 = 120$		$n_2 = 180$	
	LL	UL	LL	UL	LL	UL	LL	UL
0	0.0%	6.3%	—	—	—	—	—	—
1			0.1%	7.3%	0.1%	7.2%	0.1%	7.2%
2			0.1	7.7	0.1	7.2	0.1	7.2
3			0.3	8.4	0.2	7.3	0.2	7.2
4			0.5	9.4	0.4	7.5	0.3	7.2
5			0.9	10.5	0.6	7.8	0.5	7.3
6			1.2	11.5	0.9	8.3	0.7	7.3
7			1.6	12.6	1.1	8.9	0.9	7.5
8			2.1	13.6	1.4	9.5	1.1	7.8
9			2.5	14.6	1.7	10.2	1.3	8.1
10			3.0	15.6	2.0	10.9	1.6	8.5
11			3.5	16.7	2.4	11.5	1.8	9.0
12			4.0	17.7	2.7	12.2	2.1	9.4
13			4.5	18.7	3.1	12.9	2.3	9.9
14			5.0	19.8	3.4	13.5	2.6	10.4
15			5.5	20.8	3.8	14.2	2.9	10.9
16			6.1	21.4	4.1	14.8	3.1	11.4
17			6.6	22.3	4.5	15.5	3.4	11.9
18			7.2	23.2	4.9	16.1	3.7	12.4
19			7.8	24.1	5.3	16.8	4.0	12.8
20			8.3	25.0	5.7	17.4	4.3	13.3
21			8.9	25.9	6.1	18.0	4.6	13.8
22			9.5	26.8	6.5	18.2	4.9	14.3
23			10.1	27.7	6.9	18.7	5.2	14.8
24			10.7	28.6	6.9	19.3	5.5	15.3
25			11.3	29.5	7.3	19.9	5.8	15.7

*The two-step (double) sample confidence limits in this table are based on an initial sample of size n_1 with an additional sample of size n_2 *only* if occurrences are found in the initial sample.

Table A-2
Two-Step Sampling Confidence Limits*
Initial Sample (n_1) ***= 70***
Confidence Level = 99% (Two-Sided), 99.5% (One-Sided)
N = ***10,000***

Occurrences in Sample	Single Sample n_1		Two-Step (Double) Sample $(n_1 + n_2)$					
			$n_2 = 60$		$n_2 = 120$		$n_2 = 180$	
	LL	UL	LL	UL	LL	UL	LL	UL
0	0.0%	6.3%	—	—	—	—	—	—
1			0.1%	7.3%	0.1%	7.3%	0.1%	7.3%
2			0.1	7.7	0.1	7.3	0.1	7.3
3			0.3	8.5	0.2	7.3	0.2	7.3
4			0.5	9.5	0.4	7.5	0.3	7.3
5			0.9	10.5	0.6	7.9	0.5	7.3
6			1.2	11.6	0.8	8.4	0.6	7.4
7			1.6	12.6	1.1	8.9	0.8	7.5
8			2.0	13.7	1.4	9.6	1.1	7.8
9			2.5	14.7	1.7	10.2	1.3	8.2
10			2.9	15.7	2.0	10.9	1.5	8.6
11			3.4	16.7	2.3	11.6	1.8	9.0
12			3.9	17.7	2.7	12.2	2.0	9.5
13			4.4	18.7	3.0	12.9	2.3	10.0
14			5.0	19.6	3.4	13.6	2.6	10.5
15			5.5	20.5	3.7	14.3	2.8	10.9
16			6.0	21.4	4.1	14.9	3.1	11.4
17			6.6	22.4	4.5	15.6	3.4	11.9
18			7.1	23.3	4.8	16.2	3.7	12.4
19			7.7	24.2	5.2	16.8	4.0	12.9
20			8.3	25.1	5.6	17.5	4.3	13.4
21			8.9	26.0	6.0	18.1	4.6	13.9
22			9.5	26.9	6.4	18.7	4.8	14.6
23			10.1	27.8	6.8	19.4	5.2	14.8
24			10.7	28.6	7.2	20.0	5.5	15.3
25			11.3	29.5	7.6	20.6	5.8	15.8

*The two-step (double) sample confidence limits in this table are based on an initial sample of size n_1 with an additional sample of size n_2 *only* if occurrences are found in the initial sample.

Table A-2
Two-Step Sampling Confidence Limits*
Initial Sample (n_1) ***= 80***
Confidence Level = 99% (Two-Sided), 99.5% (One-Sided)
N = ***1000***

Occurrences in Sample	Single Sample n_1		Two-Step (Double) Sample $(n_1 + n_2)$					
			$n_2 = 60$		$n_2 = 120$		$n_2 = 180$	
	LL	UL	LL	UL	LL	UL	LL	UL
0	0.0%	5.1%	—	—	—	—	—	—
1			0.1%	6.2%	0.1%	6.2%	0.1%	6.2%
2			0.2	6.6	0.2	6.2	0.2	6.2
3			0.3	7.4	0.3	6.2	0.3	6.2
4			0.6	8.4	0.4	6.4	0.4	6.2
5			0.9	9.4	0.7	6.8	0.6	6.2
6			1.3	10.3	1.0	7.4	0.8	6.3
7			1.7	11.3	1.2	8.0	1.0	6.5
8			2.1	12.3	1.5	8.6	1.2	6.8
9			2.5	13.2	1.8	9.2	1.5	7.2
10			3.0	14.1	2.2	9.9	1.7	7.6
11			3.4	14.9	2.5	10.5	2.0	8.1
12			3.9	15.8	2.8	11.1	2.3	8.5
13			4.4	16.7	3.2	11.7	2.5	9.0
14			4.9	17.6	3.5	12.4	2.8	9.5
15			5.4	18.6	3.9	13.0	3.1	9.9
16			5.9	19.4	4.2	13.6	3.4	10.4
17			6.5	20.3	4.6	14.2	3.6	10.9
18			7.0	21.1	5.0	14.8	3.9	11.3
19			7.5	22.0	5.4	15.4	4.2	11.8
20			8.1	22.8	5.7	16.0	4.5	12.2
21			8.6	23.6	6.1	16.6	4.8	12.7
22			9.2	24.6	6.5	17.2	5.1	13.2
23			9.7	25.3	6.9	17.8	5.4	13.6
24			10.3	26.1	7.3	18.4	5.7	14.1
25			10.9	26.9	7.7	18.9	6.0	14.5

*The two-step (double) sample confidence limits in this table are based on an initial sample of size n_1 with an additional sample of size n_2 *only* if occurrences are found in the initial sample.

Table A-2
*Two-Step Sampling Confidence Limits**
Initial Sample (n_1) *= 80*
Confidence Level = 99% (Two-Sided), 99.5% (One-Sided)
N = *5000*

Occurrences in Sample	Single Sample n_1		Two-Step (Double) Sample ($n_1 + n_2$)					
			$n_2 = 60$		$n_2 = 120$		$n_2 = 180$	
	LL	UL	LL	UL	LL	UL	LL	UL
0	0.0%	5.5%	—	—	—	—	—	—
1			0.1%	6.5%	0.1%	6.4%	0.1%	6.4%
2			0.1	6.9	0.1	6.4	0.1	6.4
3			0.3	7.7	0.2	6.5	0.2	6.4
4			0.5	8.7	0.4	6.8	0.3	6.4
5			0.8	9.7	0.6	7.2	0.5	6.4
6			1.1	10.7	0.8	7.8	0.6	6.6
7			1.5	11.7	1.1	8.4	0.8	6.9
8			1.9	12.7	1.3	9.0	1.0	7.2
9			2.3	13.6	1.6	9.7	1.3	7.6
10			2.8	14.6	1.9	10.3	1.5	8.1
11			3.2	15.4	2.2	11.0	1.7	8.5
12			3.7	16.3	2.6	11.6	2.0	9.0
13			4.1	17.2	2.9	12.2	2.2	9.5
14			4.6	18.1	3.2	12.9	2.5	10.0
15			5.1	19.1	3.6	13.5	2.8	10.5
16			5.6	19.9	3.9	14.1	3.0	10.9
17			6.1	20.8	4.3	14.7	3.3	11.4
18			6.7	21.6	4.6	15.4	3.6	11.9
19			7.2	22.5	5.0	16.0	3.9	12.4
20			7.7	23.3	5.4	16.6	4.1	12.8
21			8.3	24.2	5.8	17.2	4.4	13.3
22			8.8	25.0	6.1	17.8	4.7	13.8
23			9.4	25.8	6.5	18.4	5.0	14.2
24			9.9	26.7	6.9	18.9	5.3	14.7
25			10.5	27.5	7.3	19.5	5.6	15.1

*The two-step (double) sample confidence limits in this table are based on an initial sample of size n_1 with an additional sample of size n_2 *only* if occurrences are found in the initial sample.

Table A-2

Two-Step Sampling Confidence Limits*

Initial Sample (n_1) ***= 80***

Confidence Level = 99% (Two-Sided), 99.5% (One-Sided)

N = ***10,000***

Occurrences in Sample	Single Sample n_1		Two-Step (Double) Sample $(n_1 + n_2)$					
			$n_2 = 60$		$n_2 = 120$		$n_2 = 180$	
	LL	UL	LL	UL	LL	UL	LL	UL
0	0.0%	5.6%	—	—	—	—	—	—
1			0.1%	6.5%	0.1%	6.4%	0.1%	6.4%
2			0.1	7.0	0.1	6.4	0.1	6.4
3			0.2	7.8	0.2	6.5	0.1	6.4
4			0.5	8.8	0.4	6.8	0.3	6.4
5			0.8	9.8	0.6	7.2	0.4	6.5
6			1.1	10.6	0.8	7.8	0.6	6.6
7			1.5	11.8	1.0	8.4	0.8	6.9
8			1.9	12.7	1.3	9.1	1.0	7.3
9			2.3	13.7	1.6	9.7	1.2	7.7
10			2.7	14.6	1.9	10.4	1.5	8.1
11			3.2	15.5	2.2	11.0	1.7	8.6
12			3.6	16.4	2.5	11.7	2.0	9.1
13			4.1	17.3	2.9	12.3	2.2	9.6
14			4.6	18.2	3.2	12.9	2.5	10.0
15			5.1	19.1	3.5	13.6	2.7	10.5
16			5.6	20.0	3.9	14.2	3.0	11.0
17			6.1	20.9	4.2	14.8	3.3	11.5
18			6.6	21.7	4.6	15.4	3.5	11.9
19			7.1	22.6	5.0	16.0	3.8	12.4
20			7.7	23.4	5.3	16.6	4.1	12.9
21			8.2	24.2	5.7	17.2	4.4	13.4
22			8.8	25.1	6.1	17.8	4.7	13.8
23			9.3	25.9	6.5	18.4	5.0	14.3
24			9.9	26.7	6.8	19.0	5.2	14.8
25			10.4	27.5	7.6	20.2	5.5	15.2

*The two-step (double) sample confidence limits in this table are based on an initial sample of size n_1 with an additional sample of size n_2 *only* if occurrences are found in the initial sample.

Table A-2
Two-Step Sampling Confidence Limits*
Initial Sample (n_1) = 90
Confidence Level = 99% (Two-Sided), 99.5% (One-Sided)
N = **1000**

Occurrences in Sample	Single Sample n_1		Two-Step (Double) Sample ($n_1 + n_2$)					
			$n_2 = 60$		$n_2 = 120$		$n_2 = 180$	
	LL	UL	LL	UL	LL	UL	LL	UL
0	0.0%	5.0%	—	—	—	—	—	—
1			0.1%	5.5%	0.1%	5.5%	0.1%	5.5%
2			0.2	6.0	0.2	5.5	0.2	5.5
3			0.3	6.9	0.3	5.6	0.3	5.5
4			0.6	7.8	0.4	5.9	0.4	5.5
5			0.9	8.7	0.7	6.4	0.5	5.6
6			1.2	9.6	0.9	6.9	0.8	5.7
7			1.6	10.6	1.2	7.5	1.0	6.0
8			2.0	11.4	1.5	8.1	1.2	6.4
9			2.4	12.3	1.8	8.7	1.4	6.8
10			2.8	13.2	2.1	9.4	1.7	7.3
11			3.2	14.0	2.4	10.0	1.9	7.7
12			3.7	14.9	2.7	10.6	2.2	8.2
13			4.1	15.7	3.0	11.2	2.4	8.6
14			4.6	16.5	3.4	11.8	2.7	9.1
15			5.1	17.3	3.7	12.4	3.0	9.5
16			5.6	18.2	4.1	12.9	3.2	10.0
17			6.0	19.0	4.4	13.5	3.5	10.4
18			6.5	19.8	4.8	14.1	3.8	10.9
19			7.0	20.5	5.1	14.7	4.1	11.3
20			7.6	21.3	5.5	15.2	4.4	11.8
21			8.1	22.1	5.9	15.8	4.7	12.2
22			8.6	22.9	6.2	16.4	4.9	12.6
23			9.1	23.6	6.6	16.9	5.2	13.1
24			9.6	24.4	7.0	17.5	5.5	13.5
25			10.2	25.2	7.4	18.0	5.8	14.0

*The two-step (double) sample confidence limits in this table are based on an initial sample of size n_1 with an additional sample of size n_2 *only* if occurrences are found in the initial sample.

Table A-2
Two-Step Sampling Confidence Limits*
Initial Sample (n_1) ***= 90***
Confidence Level = 99% (Two-Sided), 99.5% (One-Sided)
N = ***5000***

Occurrences in Sample	Single Sample n_1		Two-Step (Double) Sample $(n_1 + n_2)$					
			$n_2 = 60$		$n_2 = 120$		$n_2 = 180$	
	LL	UL	LL	UL	LL	UL	LL	UL
0	0.0%	4.9%	—	—	—	—	—	—
1			0.1%	5.8%	0.1%	5.7%	0.1%	5.7%
2			0.1	6.3	0.1	5.7	0.1	5.7
3			0.2	7.2	0.2	5.9	0.1	5.7
4			0.5	8.1	0.3	6.2	0.3	5.7
5			0.8	9.1	0.5	6.7	0.4	5.9
6			1.1	10.0	0.8	7.3	0.6	6.1
7			1.4	10.9	1.0	7.9	0.8	6.4
8			1.8	11.9	1.3	8.6	1.0	6.8
9			2.2	12.7	1.6	9.2	1.2	7.3
10			2.6	13.6	1.8	9.8	1.4	7.7
11			3.0	14.5	2.1	10.4	1.7	8.2
12			3.4	15.3	2.4	11.1	1.9	8.7
13			3.9	16.1	2.8	11.7	2.2	9.1
14			4.3	16.9	3.1	12.3	2.4	9.6
15			4.8	17.8	3.4	12.9	2.7	10.1
16			5.2	18.7	3.7	13.5	2.9	10.5
17			5.7	19.5	4.1	14.1	3.2	11.0
18			6.2	20.3	4.4	14.6	3.4	11.4
19			6.7	21.1	4.8	15.2	3.7	11.9
20			7.2	21.9	5.1	15.8	4.0	12.4
21			7.7	22.6	5.5	16.4	4.3	12.8
22			8.2	23.4	5.8	16.9	4.5	13.3
23			8.7	24.2	6.2	17.5	4.8	13.7
24			9.2	25.0	6.6	18.1	5.1	14.1
25			9.8	25.7	6.9	18.6	5.4	14.6

*The two-step (double) sample confidence limits in this table are based on an initial sample of size n_1 with an additional sample of size n_2 *only* if occurrences are found in the initial sample.

Table A-2
Two-Step Sampling Confidence Limits*
Initial Sample (n_1) ***= 90***
Confidence Level = 99% (Two-Sided), 99.5% (One-Sided)
N = ***10,000***

Occurrences in Sample	Single Sample n_1		Two-Step (Double) Sample ($n_1 + n_2$)					
			$n_2 = 60$		$n_2 = 120$		$n_2 = 180$	
	LL	UL	LL	UL	LL	UL	LL	UL
0	0.0%	4.5%	—	—	—	—	—	—
1			0.1%	5.8%	0.1%	5.7%	0.1%	5.7%
2			0.1	6.4	0.1	5.7	0.1	5.7
3			0.2	7.2	0.2	5.9	0.1	5.7
4			0.5	8.2	0.3	6.3	0.3	5.8
5			0.7	9.1	0.5	6.8	0.4	5.9
6			1.1	10.1	0.8	7.4	0.6	6.1
7			1.4	11.0	1.0	8.0	0.8	6.5
8			1.8	11.9	1.3	8.6	1.0	6.9
9			2.1	12.8	1.5	9.2	1.2	7.3
10			2.5	13.7	1.8	9.9	1.4	7.8
11			3.0	14.5	2.1	10.5	1.6	8.2
12			3.4	15.3	2.4	11.1	1.9	8.7
13			3.8	16.2	2.7	11.7	2.1	9.2
14			4.3	17.0	3.0	12.3	2.4	9.7
15			4.7	17.9	3.4	12.9	2.6	10.1
16			5.2	18.7	3.7	13.5	2.9	10.6
17			5.7	19.5	4.0	14.1	3.1	11.1
18			6.2	20.3	4.4	14.7	3.4	11.5
19			6.7	21.1	4.7	15.3	3.7	12.0
20			7.2	21.9	5.1	15.9	3.9	12.4
21			7.7	22.7	5.4	16.4	4.2	12.9
22			8.2	23.5	5.8	17.0	4.5	13.3
23			8.7	24.3	6.1	17.6	4.8	13.8
24			9.2	25.0	6.5	18.1	5.0	14.2
25			9.7	25.8	6.9	18.7	5.3	14.7

*The two-step (double) sample confidence limits in this table are based on an initial sample of size n_1 with an additional sample of size n_2 *only* if occurrences are found in the initial sample.

Table A-2

Two-Step Sampling Confidence Limits*

***Initial Sample* (n₁) = 100**

Confidence Level = 99% (Two-Sided), 99.5% (One-Sided)

N = ***1000***

Occurrences in Sample	Single Sample n_1		Two-Step (Double) Sample ($n_1 + n_2$)					
			n_2 = 60		n_2 = 120		n_2 = 180	
	LL	UL	LL	UL	LL	UL	LL	UL
0	0.0%	4.3%	—	—	—	—	—	—
1			0.1%	5.0%	0.1%	4.9%	0.1%	4.9%
2			0.2	5.6	0.2	4.9	0.2	4.9
3			0.3	6.4	0.3	5.1	0.3	4.9
4			0.6	7.3	0.4	5.5	0.4	4.9
5			0.8	8.2	0.7	6.0	0.5	5.1
6			1.2	9.0	0.9	6.5	0.8	5.3
7			1.5	9.9	1.2	7.1	1.0	5.7
8			1.9	10.7	1.4	7.7	1.2	6.1
9			2.2	11.5	1.7	8.3	1.4	6.5
10			2.6	12.4	2.0	8.9	1.6	6.9
11			3.0	13.2	2.3	9.5	1.9	7.4
12			3.4	13.9	2.6	10.1	2.1	7.8
13			3.9	14.7	2.9	10.7	2.4	8.3
14			4.3	15.5	3.2	11.2	2.6	8.7
15			4.8	16.3	3.6	11.8	2.9	9.2
16			5.2	17.0	3.9	12.3	3.1	9.6
17			5.7	17.8	4.2	12.9	3.4	10.0
18			6.2	18.5	4.6	13.4	3.7	10.5
19			6.6	19.3	4.9	14.0	4.0	10.9
20			7.1	20.0	5.3	14.5	4.2	11.3
21			7.6	20.7	5.6	15.1	4.5	11.8
22			8.1	21.5	6.0	15.6	4.8	12.2
23			8.6	22.2	6.3	16.1	5.1	12.6
24			9.0	22.9	6.7	16.7	5.3	13.0
25			9.5	23.6	7.0	17.2	5.6	13.4

*The two-step (double) sample confidence limits in this table are based on an initial sample of size n_1 with an additional sample of size n_2 *only* if occurrences are found in the initial sample.

Table A-2
Two-Step Sampling Confidence Limits*
Initial Sample (n_1) = 100
Confidence Level = 99% (Two-Sided), 99.5% (One-Sided)
N = 5000

Occurrences in Sample	Single Sample n_1		Two-Step (Double) Sample ($n_1 + n_2$)					
			$n_2 = 60$		$n_2 = 120$		$n_2 = 180$	
	LL	UL	LL	UL	LL	UL	LL	UL
0	0.0%	4.4%	—	—	—	—	—	—
1			0.1%	5.3%	0.1%	5.1%	0.1%	5.1%
2			0.1	5.9	0.1	5.2	0.1	5.1
3			0.2	6.7	0.2	5.4	0.1	5.1
4			0.4	7.6	0.3	5.8	0.3	5.2
5			0.7	8.5	0.5	6.4	0.4	5.4
6			1.0	9.4	0.7	6.9	0.6	5.7
7			1.3	10.3	1.0	7.5	0.8	6.1
8			1.7	11.1	1.2	8.2	1.0	6.5
9			2.0	12.0	1.5	8.8	1.2	7.0
10			2.4	12.8	1.8	9.4	1.4	7.4
11			2.8	13.6	2.0	10.0	1.6	7.9
12			3.2	14.4	2.3	10.6	1.8	8.3
13			3.6	15.2	2.6	11.1	2.1	8.8
14			4.0	16.0	2.9	11.7	2.3	9.2
15			4.5	16.8	3.3	12.3	2.6	9.7
16			4.9	17.5	3.6	12.9	2.8	10.2
17			5.4	18.3	3.9	13.4	3.1	10.6
18			5.8	19.1	4.2	14.0	3.3	11.2
19			6.3	19.8	4.6	14.5	3.6	11.5
20			6.7	20.5	4.9	15.1	3.8	11.9
21			7.2	21.3	5.2	15.6	4.1	12.4
22			7.7	22.0	5.6	16.2	4.4	12.8
23			8.2	22.8	5.9	16.7	4.6	13.2
24			8.6	23.5	6.3	17.3	4.9	13.6
25			9.1	24.2	6.6	17.6	5.2	14.1

*The two-step (double) sample confidence limits in this table are based on an initial sample of size n_1 with an additional sample of size n_2 *only* if occurrences are found in the initial sample.

Table A-2
Two-Step Sampling Confidence Limits*
Initial Sample (n_1) ***= 100***
Confidence Level = 99% (Two-Sided), 99.5% (One-Sided)
N = ***10,000***

Occurrences in Sample	Single Sample n_1		Two-Step (Double) Sample $(n_1 + n_2)$					
			$n_2 = 60$		$n_2 = 120$		$n_2 = 180$	
	LL	UL	LL	UL	LL	UL	LL	UL
0	0.0%	4.5%	—	—	—	—	—	—
1			0.1%	5.3%	0.1%	5.1%	0.1%	5.1%
2			0.1	5.9	0.1	5.2	0.1	5.1
3			0.2	6.7	0.2	5.4	0.1	5.2
4			0.4	7.7	0.3	5.9	0.3	5.3
5			0.7	8.6	0.5	6.4	0.4	5.5
6			1.0	9.5	0.7	7.0	0.6	5.8
7			1.3	10.3	1.0	7.6	0.8	6.1
8			1.6	11.2	1.2	8.2	0.9	6.6
9			2.0	12.0	1.5	8.8	1.2	7.0
10			2.4	12.8	1.7	9.4	1.4	7.5
11			2.8	13.7	2.0	10.0	1.6	7.9
12			3.2	14.5	2.3	10.6	1.8	8.4
13			3.6	15.3	2.6	11.2	2.0	8.9
14			4.0	16.1	2.9	11.8	2.3	9.3
15			4.4	16.8	3.2	12.4	2.5	9.8
16			4.9	17.6	3.5	12.9	2.8	10.2
17			5.3	18.4	3.9	13.5	3.0	10.7
18			5.8	19.1	4.2	14.1	3.3	11.1
19			6.2	19.9	4.5	14.6	3.5	11.6
20			6.7	20.6	4.8	15.2	3.8	12.0
21			7.2	21.4	5.2	15.7	4.1	12.4
22			7.6	22.1	5.5	16.3	4.3	12.9
23			8.1	22.8	5.9	16.8	4.6	13.3
24			8.6	23.5	6.2	17.3	4.9	13.7
25			9.1	24.3	6.6	17.9	5.1	14.2

*The two-step (double) sample confidence limits in this table are based on an initial sample of size n_1 with an additional sample of size n_2 *only* if occurrences are found in the initial sample.

Technical Appendix 1

Computation of Confidence Limits for Two-Step Samples

As in the computation of confidence limits for single samples, the determination of the limits for two-step samples is based on the concept of the sampling distribution.

To establish the upper limit of the confidence interval at a given confidence level (for a given sample and population size), the sampling distribution for a population with a rate of occurrence such that $(1 - CL_2)/2$ percent of the sample results will contain as few as the number actually included in the sample or less, must be found. That population rate of occurrence becomes the upper limit.

For the lower limit, the population rate of occurrence which will produce a sampling distribution with $(1 - CL_2)/2$ percent of the samples producing as many or more occurrences than that actually found in the sample, must be established.

The sampling distribution, however, will be a different form than the hypergeometric (or for large populations the binomial distribution) used in the single-sample evaluation.

The computation may be accomplished by finding the value of p (population rate of occurrence) which will result in a probability equal to $\alpha = (1 - CL_2)/2$ of producing samples with d or less for the upper limit or d or more for the lower limit.

For the upper limit, for instance, the probability that zero errors will be found in the initial sample, plus the probability that d or less will be found in the first and second samples combined after eliminating the probability of all combinations of first and second sample size which involve zero occurrences in the first sample, meets this requirement.

The formulas involved are shown below. Let:

n_1 = initial sample size,
n_2 = incremental sample size
$n = n_1 + n_2$ (combined sample size)

$$\alpha = \frac{1 - CL_2}{2}$$

where CL_2 = two-sided confidence level
0 = zero
Pr = probability
N = population size
D = number of occurrences in population,

$P = \frac{D}{N}$ (percent of occurrences in population)
d = number of occurrences found in sample
$Pr(p,n,i)$ = probability of precisely i in sample of size n

Then:

$$Pr(p,n,i) = \frac{C^i_d C^{n-i}_{N-d}}{C^n_N}$$

and generally,

$$C^a_b = \frac{b!}{a!(b-a)!}$$

For upper limit—d or less in sample:

$$\alpha = \sum_{i=0}^{i=d} [Pr(p,n,i)] + [Pr(p,n_1,0)] - [Pr(p,n_1,0)] \sum_{i=0}^{i=d} [Pr(p,n_2,i)]$$

Thus,

$$\alpha = \sum_{i=0}^{i=d} \left\{ [Pr(p,n,i)] + [Pr(p,n_1,0)] [1 - \sum_{i=0}^{i=d} [Pr(p,n_2,i)] \right\}$$

For lower limit—d or more in sample:

$$\begin{aligned} \alpha &= \frac{1 - CL_2}{2} \\ &= \sum_{i=d}^{i=n} [Pr(p,n,i)] - [Pr(p,n_1,0)] \left\{ \sum_{i=d}^{i=n} [Pr(p,n_2,i)] \right\} \\ &= \left\{ 1 - \sum_{i=0}^{i=d-1} [Pr(p,n,i)] \right\} - [Pr(p,n_1,0] \left\{ 1 - \sum_{i=0}^{i=d-1} [Pr(p,n_2,i)] \right\} \end{aligned}$$

METHOD OF COMPUTATION

By iteration find the value of p which will fulfill the stated equality (will produce probability equal to α) for each limit where for the lower limit the probability is that of finding d or more in the sample and for the upper limit finding d or less in sample. The value of p (as a percentage, for example, 7.6%) which fulfills equality for specified values of n_1 and n_2, N and confidence level is determined.

Technical Appendix 2

Two-Step Sampling vs. Single-Sampling Risks

For an attributes sample, two-sided confidence limits are established by determining the rate of occurrence p in the population which would provide a probability of

$$\alpha = \frac{1 - CL_2}{2} \qquad \text{where } CL_2 = \text{two-sided confidence level}$$

that a sample would contain as few or fewer than the number observed (d) in the sample, for the upper limit, or alternatively, as many or more than the number observed for the lower limit. This probability is the risk that the population p is outside the confidence limit *on that side*.

If the population is assumed to be very large compared to the sample size, these probabilities can be computed from the terms of the binomial distribution; if not the hypergeometric distribution must be used.

The probability of *each* possible number of sampling units of the characteristic specified being observed in the sample, when a large population contains a proportion p of sampling units of that type (such as sampling units containing errors), would then be:

$$\Pr(p,n,d) = \frac{n!}{d!(n-d)!} p^d q^{n-d}$$

where n = sample size

d = number observed in sample

p = proportion in population

$q = 1 - p$

$\Pr(p,n,d)$ = probability of d sampling units of specified type in sample of n items drawn from population containing proportion p of such units

However, in the two-step method of sampling described above, the sample is taken in two segments with a decision as to whether to continue sampling using the second segment made after the first sample has been completed. The criterion used to establish whether sampling is to be continued is whether more than some specified number of errors is found in the sample. If that number or less is found in the sample, sampling is terminated and the existing condition is assumed to be satisfactory from an audit viewpoint.

The probability that the number (d) or less will be found in the sample for a given large population and sampling will be terminated is:

$$\sum_{0}^{d} [\Pr(p,n,d)] = \sum_{0}^{d} \left[\frac{n!}{d!(n-d)!} p^d q^{n-d} \right]$$

In the most usual case, the first sample size (n_1) is chosen so that the upper limit (one-sided) will be equal to the MTER if no errors are found in the sample for a given confidence level. The risk (α_u) that the population p is actually greater than the upper one-sided limit (equal to the MTER) is then:

$$\alpha_u = 1 - CL_1 \qquad \text{where } CL_1 = \text{one-sided confidence level}$$

The probability that sampling will be terminated for a given population by finding zero errors in a sample from a large population is generally

$$\Pr(p,n_1,0) = \frac{n_1!}{0!(n_1 - 0)!} p^0 q^{n_1}$$

which resolves into for a large population[1]

$$\Pr(p,n_1,0) = q^{n_1} = (1 - p)^{n_1}$$

However, the sample size and value of p is chosen so that

$$\alpha_u = (1 - p)^{n_1}$$

[1]For a limited population,

$$\Pr(p,n_1,0) = \frac{C_0^D C_{n-0}^{N-D}}{C_n^N}$$

where D = number of occurrences in population
N = population size
n = sample size
0 = zero
C_n^N = combinations of N items taken n at a time
C_0^D = combinations of D items taken zero at a time
C_{n-D}^{N-D} = combinations of $N - D$ items taken n at a time

where p = proportion in the population when the one-sided upper confidence limit equals the MTER.

The one-sided upper confidence limit at the 95% confidence level for various-sized samples from a very large population is shown below:

Table T2-1 ***One-Sided Upper Confidence Limits for Various-Sized Samples, for a Population of 10,000 or More When Sample Contains No Errors***
95% One-Sided Confidence Level
MTER = 3%

Sample Size	Upper Confidence Limit, %
60	4.9
70	4.2
80	3.7
90	3.3
100	3.0

NOTE: Any two of the following—sample size, confidence level, and MTER—will determine the other.

As an example, using Table T2-1 at the 95% confidence level, an upper limit of 3.0% is provided when zero occurrences are found in the sample and can be used for an MTER of 3%.

It is to be noted that the above sampling plan is $n_1 = 100$, CL = 95%, with an acceptance number = zero. Unless the population contains either no errors at all or virtually none (considerably less than 1.0%) the sampling in the majority of tests will *not* terminate with the step 1 sample. See Table T2-2.

It is to be emphasized that the choice of zero as the maximum acceptance number in the step 1 sample is *not* a requirement and as discussed later may be inadvisable in many, if not most, cases.

In the two-step sampling approach, if sampling is not terminated with the step 1 sample, a second sample (n_2) is taken and a two-sided interval estimate is established.

The two-sided confidence interval for the combined sample ($n_1 + n_2$) is not equivalent to that obtained from the appropriate tables for sample size $n = n_1 + n_2$. The reason for this is shown below.

As previously, the upper limit for a sample containing d occurrences is obtained by establishing the rate of occurrence in a population which, for a given sample size, would yield a probability of α_u or (1 −

Table T2-2 ***Probability of Termination On Completion of Step 1 for Sample Size of 100 for a Population Size of 10,000***
Upper Confidence Limit = 3.0%
MTER = 3.0%

Actual Population Rate, %	Probability, %, of Termination* of Sampling on Completion of Step 1 Sample of 100
0.0	100.0
0.5	60.4
1.0	36.4
2.0	13.1
3.0	4.6
4.0	1.7
5.0	0.6

*Probability of zero in sample size used in step 1 sample.

$CL_2)/2$ for d or less occurrence in the sample. This can be calculated from

$$\alpha_u = \sum_{i=0}^{i-d} [\Pr(p,n,i)]$$

If the upper limit is computed on the basis of the total sample without regard to the two-step process (as a single sample of size $n_1 + n_2$), then

$$n = n_1 + n_2$$

where n_1 = first-step sample size
n_2 = second-step sample size
α_u = upper-limit risk $\frac{1 - CL_2}{2}$

However, for some samples, sampling will have been terminated upon completion of the first-step sample using only n_1 sampling units. The risk $(1 - CL)$ in that first sample is α_1. The risk $(1 - CL)$ on the step 2 sample (α_2) must be considered as beyond and in addition to the risk of being wrong (α_1) of the first-step sample.[2]

$$\alpha = \alpha_1 + \alpha_2$$

to obtain the risk of being wrong on either step.

[2]These risk probabilities may be added since they are mutually exclusive.

The *upper-limit* risk is the probability that, although the error rate is at the MTER, either the first-step sample will produce no errors and thus find the population satisfactory or on the whole sample the actual error rate will exceed the upper limit (MTER).

The actual overall risk (probability of being wrong) α for the upper limit for a sample of a given size containing d occurrences can be computed indirectly from the following facts.

If the $n = n_1 + n_2$ items were taken as a *single sample* with an upper limit such that $\alpha_u = (1 - CL_2)/2$, then

$$\alpha_u = \sum_{i=0}^{i=d} [\Pr(p,n,i)]$$

For a two-step sample, the risk of a false decision (α) on the first-step sample is the probability that zero occurrences will be found even though the error rate equals the MTER. This is equal to

$$\alpha = \Pr(p,n_1,0)$$

when p = rate of occurrence in population equal to upper limit (MTER). Therefore,

$$\alpha = \sum_{i=0}^{i=d} [\Pr(p,n_2,i)] + [\Pr(p,n_1,0)]$$

However, if sampling is terminated when zero is found in the first sample, there is no possibility of zero in the overall sample when sampling is completed for the other samples nor can any combination of zero in the first sample and anything else in the second sample take place. The probability that such combinations would occur had sampling not been terminated is equal to the product of their probabilities, and hence for all possible combinations of zero in the first step and zero to d in the second sample, the total probability will be

$$\sum_{i=0}^{i=d} [\Pr(p,n_1,0)]\,[\Pr(p,n_2,i)]$$

or

$$\Pr(p,n_1,0) \sum_{i=0}^{i=d} [\Pr(p,n_2,i)]$$

Since these events were included in the probability (α) associated with the overall ($n = n_1 + n_2$) sample, these probabilities must be deducted.

Hence, for the upper limit

$$\alpha_u = \sum_{i=0}^{i=d} [\Pr(p,n,i)] + [\Pr(p,n,0)] - [\Pr(p,n,0)] \sum_{i=0}^{i=d} [\Pr(p,n_2,i)]$$

It is possible to select a value of p, n, and d so that α (or the risk) will provide a desired confidence level but it is quite clear that this will not be equal to the confidence level originally selected for the first-step sample.

In other words, it is possible to select a two-step sampling plan so as to provide a given confidence level for the final overall sample appraisal but it will not be possible to have the same confidence level as that for the first step sample. Nor will the limits so established be the same as that for the overall sample taken in one step.

The higher overall risk occurs in spite of the fact that the formula shows both an addition and subtraction. With the exception of the limited case where $\Pr(p,n_1,0)$ equals zero the added component will always be larger than the subtracted component, since

$$\Pr(p,n_1,0) > \Pr(p,n_1,0) \sum_{i=0}^{i=d} [\Pr(p,n_2,i)]$$

Where $\Pr(p,n_1,0)$ equals zero, the two risks will be equal.

Therefore, for a given p in the population and given sample size the risk α_u for a two-step sample will be greater than for a single sample of size $n_1 + n_2$.

Alternatively, for a specified equivalent risk (or confidence level) for a given sample size $(n_1 + n_2)$ taken as a two-step sample rather than a single sample, the confidence interval estimate will be wider. However, as the probability of zero in the first sample approaches zero, the risks for a given sample size for the two methods will tend to approach each other.

The probability of zero in the first sample will approach zero as the rate of occurrence (p) in the population and the sample size (n_1) grows.

As a result, the confidence limits for the two types of sampling for a given final sample size will tend toward each other as these values increase.

Technical Appendix 3

AICPA Formula for Selecting a Confidence Level Based on Compliance Tests Results

It is apparent that the formula given in *SAS 1* for determining the confidence level for a substantive test based on the outcome of the audit examination of compliance with internal controls,

$$S = 1 - \frac{(1 - R)}{(1 - C)}$$

where S = confidence (reliability) level for a substantive test
R = combined confidence (reliability) level desired
C = reliance assigned to internal accounting control and other relevant factors based on auditor's judgment

is an attempt to develop an approach which is based on accepted probability theory.

If the intent was to provide a basis in probability theory, the formula must have been created in terms of either (1) the classical approach or (2) the Bayesian approach.

If the classical approach was intended, the method must rely on the well-known multiplication rule. The rule states that if several *independent* chance events can occur (e.g., a head on tossing a coin and a black card drawn from a deck of cards), the probability that they will all occur is equal to the products of their several probabilities.

Thus, in the present situation, the *risk* in the compliance examination (failure to find that the internal controls are not applied properly) multiplied by the probability of failure to detect a material error by the substantive test equals the overall probability that the material error will not be detected. However, there are some deficiencies in this approach:

1. The two events are not of necessity independent.
2. The value of the "auditor's judgment concerning the reliance to be assigned to internal control and other relative factors" seems to be a subjective *rating* rather than an objective probability.
3. The risk for the substantive test (1 − CL) does *not* indicate the risk of being wrong about the conclusion that no material error exists. A confidence level, from which such a risk is determined, expresses the probability that the true audited value is included within the confidence interval.

   If, in fact, the audited population value is outside the limits but, nevertheless, the book value falls inside the limit, this circumstance merely indicates that there is an undetected statistically significant difference between the book value and the "true" value. However, such a difference *may or may not* be material.
4. The overall risk (1 minus the "combined reliability level desired") is specified as a reliability (confidence) level and seems to be equated to the risk of failing to detect a material error. The two are not the same.

If the Bayesian approach was intended, independence is not required and the formula becomes

$$\Pr(C' \text{ and } S') = [\Pr(C')]\left[\Pr\left(\frac{C'}{S'}\right)\right]$$

where the overall probability of failure of both the compliance and substantive test is equal to the probability of failure of the compliance examination multiplied by the probability of failure of the substantive test.

No independence is required in this formula but questions again can be raised. Is the "auditors judgment concerning reliance to be assigned to internal accounting control and other relevant factors" a probability value or merely a rating? If a probability figure is intended, what probability outcome is intended? Is it the probability that the compliance examination will fail to detect a lack of compliance which might result in failure to prevent a material error or is it a statistical probability (the confidence level) that the population audited total is included between the confidence limits? *SAS 1* does not indicate which probability is intended but its use by some seems to indicate that it is interpreted as the confidence level in a compliance test. The "combined reliability level" is similarly ill-defined.

In addition, there is a general confusion between statistical risks and audit risks, which are not equivalent. (See Chapter 10 for a discussion of these risks.)

There is no generally accepted statistical basis for using this method to establish confidence levels for substantive tests. Until some rationale based on sound probability principles is provided, the use of this approach to establish confidence levels is not defensible, especially because its use may result in confidence levels as low as 50% for substantive tests.

Technical Appendix 4

Measurement of Skewness and Kurtosis

The normal distribution is one of a series of distributions which are symmetrical. The portion of the distribution above the average is the same as that below but in the opposite direction.

However, there are many asymmetrical frequency distributions. Virtually all distributions of accounting data are skewed, usually positively. The degree of skewness can be measured by a value:[1]

$$\alpha_3 = \frac{\Sigma(X - \overline{X})^3}{n\sigma^3}$$

where X = each individual value
$\overline{X}$ = arithmetic mean of all values
σ = standard deviation
Σ means "the sum of"

The computation of the value is illustrated below.

The value of α_3 for all symmetrical distributions including the normal curve is zero. Refer to Figure T4-1 for an indication of the significance of different values. Values of α_3 for accounting data may be as high as 6 or 7 in extreme cases.

The extent to which values are concentrated about the average as contrasted with their spread is sometimes referred to as *peakedness*. The measure of this characteristic, the tendency for the distribution to

[1]The result of this calculation can be either positive or negative depending on the direction of the skewness.

Table T4-1

Item Number	(X) Dollar Value	$(X - \overline{X})$	$(X - \overline{X})^2$	$(X - \overline{X})^3$	$(X - \overline{X})^4$
1	$20.	0	0	0	0
2	25.	+5.	25.	125.	625.
3	10.	−10.	100.	−1,000.	10,000.
4	10.	−10.	100.	−1,000.	10,000.
5	60.	+40.	1,600.	+64,000.	2,560,000.
6	15.	−5.	25.	−125.	625.
7	30.	+10.	100.	+1,000.	10,000.
8	13.	−7.	49.	−343.	2,401.
9	12.	−8.	64.	−512.	4,096.
10	5.	−15.	225.	−3,375.	50,625.
			2,288.	58,770.	2,648,372.
Total	$200.00				

$$\overline{X} = \frac{\$200.00}{10} = \$20$$

$$\alpha = \sqrt{\frac{\Sigma(X - \overline{X})^2}{n}} = \sqrt{\frac{2288.00}{10}} = 15.1261$$

$$\alpha_3 = \frac{\Sigma(X - \overline{X})^3}{n\alpha^3} = \frac{58{,}700}{(10)(15.1261)^3} = +1.7$$

$$\alpha_4 = \frac{\Sigma(X - \overline{X})^4}{n\alpha^4} = \frac{2{,}648{,}372.00}{(10)(15.1261)^4} = 5.1$$

be relatively flat or with high concentrations about the average, called kurtosis, may be measured by

$$\alpha_4 = \frac{\Sigma(X - \overline{X})^4}{n\sigma^4}$$

For a normal distribution, the value of α_4 equals 3. The measurement of skewness is frequently expressed as $\alpha_4 = 3$ to yield a value of zero for the normal curve. Values of $\alpha_4 - 3$ for accounting data may be as high as 40 or 50 if not stratified.

Some time-sharing computer systems used in conjunction with the calcuations for sampling in auditing automatically calculate the sample values of α_3 and α_4.[2]

[2]The U.S. Department of Health, Education and Welfare (HEWCAS) Program performs this calculation for variable sample evaluations, as does the U.S. Internal Revenue Service sampling program PAL, parts II and IV.

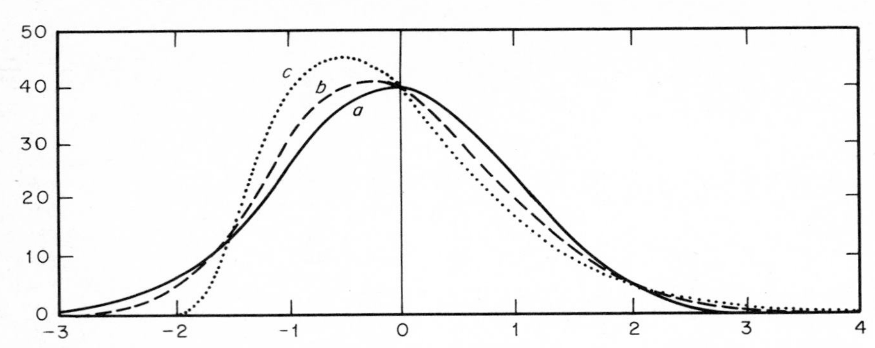

Fig. T4-1 Curves of varying degrees of skewness: (*a*) $\alpha_3 = 0$; (*b*) $\alpha_3 = .5$; (*c*) $\alpha_3 = 1.0$. SOURCE: D. J. Cowden, *Statistical Methods in Quality Control*, Prentice-Hall, Englewood Cliffs, N.J., 1957, p. 19. Used by permission.

The calculation of α_3 and α_4 is illustrated in Table T4-1 using a few simple figures. It is to be emphasized that such computations should be applied only to larger groups of data in practice. The small group is used here merely to facilitate the illustration.

The use of a small electronic calculator with a y^x key and a memory eliminates much of the detail calculation shown above, by cumulating $(X - \overline{X})$ to the desired powers without the need for preparing a table such as that shown above.

Technical Appendix 5

Confidence Limits for Averages of Samples from Nonnormal Populations

Confidence limits for interval estimates of averages (or totals) are computed by adding and subtracting the sampling error from the point estimate of the average (or total). The formula for the sampling error of an average is

$$SE_{\bar{x}} = t\frac{\sigma}{\sqrt{n}}\sqrt{1-\frac{n}{N}}$$

where t is the normal curve deviate.

The sampling error formula thus assumes that the sampling distribution of the average forms a normal distribution. If the sampling distribution is not normal, the result of the computation is not correct. Nevertheless, this formula is used regardless of the nature of the population sampled. This is called the normal approximation.

As explained in Chapter 3, regardless of the nature of the population sampled, the central-limit theorem indicates that as the sample size grows, the sampling distribution of the average will approach the normal form. For very large samples, the sampling distribution is indeed normal or very close to it. However, for samples of the size frequently used in auditing, the sampling distribution of the average will not be normal if the population is badly skewed.

Nevertheless, it is possible to estimate the confidence limits in such situations without resort to the normal approximation.

Pearson has provided tables which can be used for this purpose.[1] With reference to these tables he states:

> A considerable proportion of current statistical research is concerned with the determination of the sampling distributions of statistics required either as estimators or for use in tests of significance. It is often the case that while the distribution itself cannot be expressed in any simple form, the sampling moments can be derived and numerical values calculated either precisely or as approximations.[2]

The table provides values of t which are appropriate when the sampling distribution will not be normal for various values of the measures of skewness and kurtosis for that distribution.

It has been shown that the measure of skewness of the sampling distribution $(\alpha_{3\bar{X}})$ of averages (or totals) can be calculated from that for the population by

$$\alpha_{3\bar{X}} = \frac{\alpha_3}{\sqrt{n}}$$

where $\alpha_{3\bar{X}}$ = skewness of the sampling distribution
α_3 = skewness of population
n = sample size[3]

In a similar manner, the kurtosis of the sampling distribution of the averages can be established from the measure of kurtosis for the population by:[4]

$$(\alpha_4 - 3)_X = \frac{\alpha_4 - 3}{n}$$

where $(\alpha_4 - 3)_X$ = measure of kurtosis of sampling distribution
$\alpha_4 - 3$ = measure of kurtosis of population
n = sample size

In practice, it is necessary that the values of α_3 and α_4 be computed from the sample of audited values (or errors) since these audited population values are not known.

[1]E. S. Pearson and H. O. Hartley (eds.), *Biometrica Tables for Statisticians*, vol, 1, Cambridge University Press, New York, 1954, table 42.

[2]Ibid., p. 82.

[3]F. N. David, *Probability Theory for Statistical Methods*, Cambridge University Press, New York, 1951, pp. 131–132.

[4]Ibid.

The estimate of the α_3 and α_4 values from the sample introduces a sampling error for these estimates, but since

$$\sigma_{\alpha_3} = \sqrt{\frac{6}{n}}$$

$$\sigma_{\alpha_4} = \sqrt{\frac{24}{n}}$$

for sample sizes of 100 or more, these sampling errors are not important.

The tables on pages 67 to 69 provide these values, which are used to replace the normal curve values of t in calculating the sampling error.

The values in Table 3-6 were extracted from the Pearson table but are presented in a somewhat more useful form for the purposes of this book. The original table used β_1 and β_2 as entry values, where

$$\beta_1 = \alpha_3^2$$
$$\beta_2 = \alpha_4$$

Index

About the Author

Dr. Herbert Arkin has specialized in the application of statistical sampling techniques to the fields of auditing and accounting since 1950. He is professor emeritus of the Bernard M. Baruch College of the City University of New York, where he served on the faculty (undergraduate and graduate) since 1930. He was the head of the Statistics Department of that college and was designated Distinguished Professor.

He has written a number of books, several of which have been translated into foreign languages. The most relevent of these books to the present subject matter is *The Handbook of Sampling for Auditing and Accounting*, 2d ed. (1974). He has also written numerous articles for professional publications.

In the area of the accounting applications of sampling, Dr. Arkin has served as consultant to Price Waterhouse and Co., Arthur Young and Co., as well as many other accounting firms. He has also been a consultant to the audit staffs of numerous federal, state, and local audit agencies. He has conducted training courses in statistical sampling for auditors for a wide variety of organizations in the United States and Canada. In addition, he has served as a sampling consultant for a considerable number of business organizations in nonaccounting applications of sampling. He has served as an expert witness in statistical sampling before a variety of courts and commissions.